LINEAR PROGRAMMING APPLICATIONS TO AGRICULTURE

Linear Applications

RAYMOND R. BENEKE

Programming to Agriculture

RONALD WINTERBOER

The Iowa State University Press, AMES

1973

Raymond R. Beneke has taught and done research on the economics of agricultural production at Iowa State University for twenty-five years. He was the recipient of an outstanding teacher award by Iowa State University in 1968 and an award for professional excellence from the American Agricultural Economics Association the same year. Professor Beneke holds B.S. and M.S. degrees from Iowa State University and the Ph.D. degree from the University of Minnesota.

Ronald D. Winterboer holds B.S. and M.S. degrees from Iowa State University where he is an instructor and extension economist.

Composed and printed by
The Iowa State University Press

First edition, 1973
Second printing, 1976
Third printing, 1980

Library of Congress Cataloging in Publication Data

Beneke, Raymond R 1919–
Linear programming applications to agriculture.

1. Farm management. 2. Linear programming.
I. Winterboer, Ronald, 1946– joint author.
II. Title.
S566.B43 658.4'033 72-2298
ISBN 0-8138-1035-3

CONTENTS

PREFACE

This book is the product of six years experimentation in teaching farm planning to seniors in the College of Agriculture at Iowa State University. The emphasis throughout is on application of linear programming methods. Many excellent references are available on the theoretical structure and implications of activity analysis, but the authors are convinced that the student is rare indeed who can understand well the potential and limitations of the method without constructing, optimizing, and experimenting with a variety of models.

This book provides an approach whereby conscientious students with a modest background in algebra, economics, and computer science can achieve, in a single course, a working knowledge of linear programming. Emphasis is given to partial models, each designed to illustrate one or two ideas. All models assume a thorough knowledge of agriculture; they may have a wider application in other industries. In no case should the student assume that the models are a convenient and accurate source of coefficients. They are offered for illustrative purposes only. An alert student in agriculture should be able to improve upon them for the planning situation with which he is working.

The sections on data preparation and processing are specific to the computing facility at Iowa State University, but this system is widely used. Therefore, the material presented here should be applicable with minor modifications by anyone who has access to an IBM 360 installation.

LINEAR PROGRAMMING APPLICATIONS TO AGRICULTURE

CHAPTER ONE

BACKGROUND AND HISTORY

Linear programming is a planning method that is often helpful in decisions requiring a choice among a large number of alternatives. The theoretical concepts on which the method depends have been known for many years. However, it was during World War II and immediately after that its application to planning problems first was stressed. Since then it and similar techniques have been applied increasingly to management decisions in industry. Where should production facilities and warehouses be located in respect to sources of raw materials and markets for the finished product? What mix of ingredients will minimize the cost of producing feed, gasoline, or fertilizer? How can production be scheduled to achieve the greatest output of product from plant and equipment? These are a few of the questions programming techniques sometimes can help answer.

The method, which grew out of applied mathematics, is constantly being refined so that it can be applied with greater precision to a wider range of problems. Like many innovations, its usefulness would have been limited without a parallel technological development, the electronic computer. As we shall see later, a large volume of computations is required to plan realistically with this method.

The first efforts to apply programming to farm problems were crude and produced few useful results. Subsequent improvements in the method and in development of electronic computers and effective computing routines to guide them have made linear programming a useful tool for analyzing the optimum organization of the farm business.

WHY STUDY LINEAR PROGRAMMING?

Linear programming is sufficiently complex to require several months of intensive study and practice to master its application to farm planning. In addition, computer facilities must be available. For this reason, the process of planning farms by programming is a job for the specialist. Why then should a student with no intention of becoming a farm planning specialist study programming? The answer is that by studying the method he develops an appreciation for the complex manner in which prices; yields; and such scarce resources as land, capital, and labor interact during critical seasons to determine the best farm plan. Of equal importance is the fact that many farm management studies now use programming techniques. To understand these research reports requires knowledge of the method being used.

ADVANTAGES OF PROGRAMMING

The typical farm business has on hand or available to it a supply of labor, capital items, and land that can be allocated to the production of numerous crops and livestock products. Each input can be allocated among production possibilities in thousands of ways. The number of possible alternative plans in a farm business extends to millions because of the diverse resources used and the wide range of production alternatives that are feasible on the typical farm.

With such a huge number of alternatives to choose among, how can any manager select the best plan? Usually he is guided by what has proved successful in the past both for him and for others. He adjusts his existing program "on the margin," making relatively small changes from year to year. In planning these year-to-year adjustments he may use informal arithmetic or a more or less systematic budget approach. Although the plan which results from this process may not be the best available, typically it makes reasonably good use of the farm's land, labor, and capital. The great advantage of programming is that it allows one to test a wide range of alternative adjustments and to analyze their consequences thoroughly with a small input of managerial time.

The question, What would happen if . . . ? can be posed repeatedly and answered rigorously and quickly. Examples of questions of this type are: What would happen to expected income and the optimum mix of activities (a) if 80 acres more land were available? (b) if hogs were 20 cents per pound instead of 22 cents? (c) if the number of feeder cattle were limited to 50? (d) if another man were hired? (e) if hog farrowing and finishing capacity were increased? Thus the principal advantage of linear programming as a planning method is not that it leads to one foolproof plan but that it provides a means of analyzing a variety of alternative decisions.

INFORMATION REQUIRED FOR PROGRAMMING

The information needed to develop a farm plan by linear programming is much the same as is used in budgeting except that programming demands a more rigorous specification of restraints and more detailed input-output data. To program, one must have certain building blocks.

Activities To Be Considered in the Planning Process

Production activities are comparable to enterprises except that they are more precisely defined. A one-litter hog program with sows farrowing in March is one activity; the same program with sows farrowing in June is another. A corn-corn-oats-meadow (CCOM) rotation with no fertilizer is one activity; CCOM with fertilizer is another. If one production process differs from another in the type, proportions, or timing of the inputs it uses—i.e., has different coefficients—the two are treated as separate activities. In preparing a typical planning model, one may work with fifteen or twenty crop production activities and at least as many livestock activities.

In addition to production activities, activities such as buying and/or selling inputs and products are usually provided. Production processes likewise can be subdivided into numerous activities. In corn production, for example, seedbed preparation, planting, weed control, harvesting, and drying can each be treated as a specific activity. The number and structure of activities the planner should include in a model are always a function of the answers he is seeking. Analyses of highly complex relationships may require a sophisticated model containing hundreds of activities.

Production Coefficients

As a first approximation, the reader may look upon coefficients as corresponding to the crop yields and the feed, labor, and capital requirements used in budgeting. Production coefficients are always stated in terms of the amount of input required per unit of activity. This is the reverse of the traditional mode of specifying land coefficients in crop production where the relationship is usually framed in terms of the amount of output per acre of land.

The livestock coefficients for programming parallel the resource requirements used in budgeting. If the activity unit for swine production is defined as one litter, a programming model requires estimates of the amount of grain, pasture, labor, and capital used to produce a litter. Although the input-output coefficients needed to program are often difficult to estimate, they are no more so than for other planning methods, provided one tries to be equally as rigorous with other approaches.

Product and Input Prices

Formulating price expectations for programming is little different than for budgeting. In both, the usefulness of the plan that finally emerges is a function of the accuracy with which prices are predicted. In programming, as in budgeting, emphasis should be placed on accurate relative prices. Having all prices too high or too low will distort the net income estimate, but if prices are in line with one another the farm plan developed will still be useful. As in budgeting it is important that seasonal price variations be taken into account in formulating price expectations in the programming model. Thus the price attached to an activity which provides for farrowing pigs in February and selling them in August should reflect the price expected in August, not the expected annual mean.

Restraints

Restraints may also be referred to as restrictions; the same meaning is intended for both terms. This concept represents the major departure in terminology from that used in budgeting. In a loose way, restraints include the land, labor, and capital available or which can be made available on the farm for production purposes. In addition, restraints imposed by government programs such as acreage allotments are often essential in the planning model. The operator also may impose restrictions on himself, such as refusing to keep more than two dairy cows or to raise more than ten acres of hay. Always one and usually several restraints will limit the attainment of higher income. Likewise, no operator has access to an unlimited quantity of any resource, and hence a restraint on its use eventually will emerge. The amount of level productive land may be one restriction, the acres of rolling land a second, and the land that must be kept in permanent pasture a third. The amount of labor available each month of the year may be included in the program as twelve separate restrictions. Indeed, labor during one or two weeks of the critical planting season could constitute a separate restriction.

Maximum restraints are "no more than" restraints. The model specifies that no more than some quantity of resources may be used in production or that an activity can enter the plan at no more than some level. Planning models may also include "no less than" (minimum) restraints or "equal to" (equality) restraints.

PROGRAMMING AND FARM BUSINESS GOALS

The programming procedure is designed to specify the farm plan which will yield maximum income given the restraints, prices, and yields anticipated. This does not mean that objectives of the operator such as avoiding risk or working with enterprises that give him personal satisfaction must be ignored. If he feels that cattle feeding, lamb feeding, or turkey production are too risky, the planner can arrange to leave them out of the range of activities included in the planning model. Or if he has always had a dairy herd and would be unhappy without this enterprise, the programming model can be restrained so as to force this activity into the plan. If he is conservation-oriented and wishes to keep half his cropland in grass and legume crops, a plan meeting this condition can be arranged. Always, however, the programming process will search out the plan promising the most profit, taking into account the conditions the operator has imposed. If the operator insists on too many restrictions which limit the farming program, the optimum plan may not be a very profitable one.

LEVELS OF MANAGEMENT

We know from our observations of farm operators that some are more skillful processors of inputs into products than others. In addition, we know that one operator may do an excellent job with hogs but perform cropping operations poorly. Another may raise the best corn in the vicinity but have great difficulty in maintaining a high level of production with a dairy herd. The planner has the opportunity to take the quality of management into account by fitting the production coefficients—the amount of feed, labor, land, and capital required per unit of product—to the operator involved. If the latter can achieve yields no higher than 100 bushels of corn or 9,000 pounds of milk per cow, this level of performance should be reflected in the coefficients specified in the model.

Some managers may have the capacity to work effectively with a small enterprise but fail with a larger one. For example, an operator may be capable of managing a 30-litter hog enterprise well but have great difficulty when the number is increased to 100 litters per year. Management restraints limiting the scale of an enterprise to any level desired can be included in the model without difficulty.

LIMITATIONS IN THE USE OF LINEAR PROGRAMMING

Thus far we have emphasized the usefulness of linear programming as a farm planning method. Lest the reader infer that programming can solve all the problems associated with farm planning, we discuss below the steps in the planning process wherein programming affords no help and the areas in which its use is subject to important limitations.

1. Programming cannot help the manager in the difficult task of formulating price expectations. The process can only indicate the best way to use resources once a judgment has been made as to future prices. A plan based on prices which in retrospect prove too low will turn out to be less profitable than would one based on more accurate price expectations. This problem is not peculiar to planning with the programming technique. Any type of planning—whether done in the manager's head, by budgeting, or by programming—requires an estimate of what prices will be in the future. Serious mistakes in estimating prices, especially relative prices, will lead to poor results in any type of planning.

2. Programming is of little help in estimating input-product relationships. The method can only specify the type and quantity of data needed. The planner must supply estimates of the amount and distribution of labor, feed, land, and capital needed to produce crops and livestock. Estimates of this type are difficult to make, especially on farms where record keeping has been neglected.

3. Programming proceeds as if the price and input-output expectations we have formulated were equally reliable for all farm products. As a result, all enterprises are treated as though they were equally without risk. Thus the resulting plan does not take into account the risk preferences of the operator. However, as we noted previously, it is possible for the operator to exclude or limit the scale of enterprises he considers too risky by eliminating them from the range of activities offered or by restraining their level.

4. Restraints are sometimes difficult to specify. A plan typically looks ahead at least a year. It may be difficult to know how much labor will be available during the coming summer; the supply of hired labor may be unpredictable and the supply of family labor uncertain. Where the operator uses credit in the business, he may be uncertain as to how he will fare in his negotiations with his creditors as the farm plan is carried out.

5. One of the assumptions of linear programming is that each additional unit of output requires the same quantity of input. The student will recall that the amount of crop output per unit of fertilizer declines as more fertilizer is used per acre and that pounds of milk per unit of feed decline as a dairy cow is fed at higher levels. These are two of the many instances of diminishing marginal returns in farm production. Taking diminishing returns into account is troublesome because more activities must be included in the model. However, production relationships of this type can be handled realistically. Corn produced with a median level of fertilizer is treated as one ac-

tivity and corn with a high level as another. A choice among any number of different levels can be provided for in this way. A similar procedure can be followed wherever a choice among different levels of input is desired, provided the amount of output per unit of input decreases as the input is used more intensively.

6. Activities that involve decreasing costs cannot be treated adequately with present programming methods. This problem, although it also arises from the assumption of linear production coefficients, is the opposite of the one discussed above. In this case progressively smaller amounts of input are required per unit of activity. One example of a production relationship of this type is the declining amount of labor required per cow as the number of cows in the herd is increased. In the case of machinery, the fixed costs per acre such as depreciation, capital, and taxes decline as the number of acres of annual use made of the machinery increases. A procedure to compare the profitability of high- versus low-capacity machines in the same programming solution has not been developed. Comparisons between two or more capital expenditures involving different levels of fixed costs can be accomplished only through a separate optimization for each capital item. Thus comparing high-capacity with medium-capacity machines requires two optimizations—one for each machine combination. The net income from the two plans can then be compared. Similar investment decisions abound on the typical Corn Belt farm. Although progress is being made in developing models and computational procedures which ease the problems surrounding analysis of investment decisions, their application remains awkward.

The decrease in per unit labor requirements as the activity level increases usually does not lead to seriously distorted plans, because most enterprises enter the program, if at all, at a level where the labor requirement per unit of activity has become relatively constant. Thus we can often anticipate quite accurately what the labor requirements per cow, litter, or acre will be and use this coefficient in planning.

7. Facilities including well-tested routines as well as the computer must be available for successful application of linear programming to farm planning. Achieving realistic results with programming requires the use of a large number of activities and restrictions. As these increase, the computational task quickly becomes too burdensome for the common desk calculator and can be dealt with economically only with an electronic computer.

CHAPTER TWO

LINEAR PROGRAMMING PROCEDURES

In this chapter we present an example of a simple crop production problem. We first show how the farm planner organizes the resource and production information he obtains from the farm into a format convenient for applying the computations of the simplex method. The information contained in the problem solution is described and interpreted, completing the survey of the minimum knowledge required in linear programming applications to farm planning.

To help develop a better understanding of the method, we present the algebraic formulation of the same problem and discuss the programming computations and algebraic presentations of the solution. Finally, we present the dual solution to the same problem.

SIMPLE CROP PRODUCTION PROBLEM

The Problem

The problem is how to organize the farm business (i.e., what crops to raise) to maximize net returns over variable costs, given the conditions described below.

The restraints are:

Land .... 12 acres
Labor .... 48 hours
Capital $360

The activities are:

Corn production
Soybean production
Oat production

The activities are defined in units of one acre; i.e., one acre of corn production, one acre of soybean production, etc. The coefficients and net prices described below refer to one-acre units.

The coefficients are:

Corn production requires 1 acre of land, 6 hours of labor, and $36 of capital.
Soybean production requires 1 acre of land, 6 hours of labor, and $24 of capital.
Oat production requires 1 acre of land, 2 hours of labor, and $18 of capital.

Net Prices

The net price of an activity as defined here is the value of gross sales minus variable costs of production. A more rigorous definition will be introduced later. In our example, if a unit (one acre) of the corn production activity produces gross value of sales of (say) $75 and if variable costs of producing that acre of corn were $35, the net price would be $40.

Net prices used in this example are:

Corn production .... $40 per acre
Soybean production .... $30 per acre
Oat production .... $20 per acre

Arranging the Problem

The farm planner estimates all the above coefficients from farm records or observations of the farm operator, supplementing if necessary from his own experience as well as from data collected by other farmers and from experiment station work. He next arranges the data into a particular format, or matrix. In doing so, he first lists the restraints and their appropriate levels for the planning situation under study. The production activities are arranged as a series of columns. A coefficient at the intersection of a column and row shows how much of the resource in that row is used by one unit of the production activity. The format that results is shown in Table 2.1.

For example, producing one unit (1 acre) of corn requires 1 acre of land, 6 hours of labor, and $36 of capital. The net price (gross sales minus variable costs) is $40 per acre.

TABLE 2.1: Crop Production Problem Arranged in Matrix Format

		Production Activities			Disposal Activities		
Restriction	Restriction Level	Corn (1 acre)	Soybeans (1 acre)	Oats (1 acre)	Land	Labor	Capital
Land	12 acres	1	1	1	1		
Labor	48 hours	6	6	2		1	
Capital	$360	36	24	18			1
Net Price		40	30	20			

Disposal or Slack Activities

Disposal activities are included to allow nonuse of resources. They also are referred to frequently as slack activities or slack variables. Their formulation will be discussed more fully later; it is sufficient here only to indicate their purpose. Since the income maximizing farm plan realistically need not use the entire quantity of each resource available, provisions must be made for the nonuse of resources in the final plan. The disposal or slack activities allow this nonuse to occur.

INFORMATION CONTAINED IN THE SOLUTION

The solution typically contains three classes of information useful to the farm planner. This information is described briefly in the following sections.

Value of the Program

The value of the program resulting from the optimization of the model presented in Table 2.1 is the total gross sales from all production activities included in the final plan, minus their variable costs. Subtracting all fixed costs from the value of the program would give net farm income. In this example, fixed costs must be subtracted, since the net activity prices used in the planning model accounted only for variable costs.

The plan that maximizes the value of the program also maximizes net farm income and labor income for the resources used. In our crop production example, the value of the program turns out to be $360 in the final plan. There is no way that the farm business can be reorganized, given the resource restraints and price relationships assumed in the problem, to yield a value of the program greater than $360.

Final Plan

Our objective was to organize the farm business so that net returns over variable costs would be maximized, given the resources available and the net prices used. In the crop production problem, the final (optimal) plan included:

Corn production .... 6 acres
Oat production .... 6 acres
Unused capital $36

There is no way that the land, labor, and capital available can be recombined that will generate more income than the above combination.

Shadow Prices

Shadow prices for production activities indicate how the value of the program would change (how much income would be penalized) if an additional unit of the activity were forced into the final plan. They will be referred to frequently in this book as income penalties.

In our example problem, soybeans did not enter the final plan. If an acre were for some reason "forced in" the final plan (replacing an acre of corn), the value of the program would be reduced by $10, from $360 to $350.

Shadow prices for the disposal activities provide information concerning the productivity of added resources (relaxation of restraints). All the land available in our problem was planted to corn and oats. An additional acre, if made available, would add $10 to the value of the program. (More precisely, if one acre of land were taken away, the value of the program would be reduced by $10.) An additional hour of labor would add $5 to the value of the program, but more capital would add nothing since the original supply was not fully used in the final plan.

ALGEBRAIC FORMULATION

The crop production problem can also be stated algebraically. Let:

X_C = units of corn produced
X_S = units of soybeans produced
X_O = units of oats produced

The units of activity, in this case production of crops, are acres.

Next we form a series of inequalities which define the conditions within which we seek to maximize profits:

$$1X_C + 1X_S + 1X_O \leq 12 \text{ acres of land} \quad (2.1)$$
$$6X_C + 6X_S + 2X_O \leq 48 \text{ hour of labor} \quad (2.2)$$
$$36X_C + 24X_S + 18X_O \leq \$360 \text{ capital} \quad (2.3)$$

The second of the inequalities above, for example, specifies that 6 hours of labor times the acres of corn produced plus 6 hours of labor times the acres of soybeans produced plus 2 hours of labor times the acres of oats produced must be less than or equal to 48 hours, the total amount of labor available. The first and third inequalities give similar conditions for land and capital. The total quantity of land used must be less than or equal to 12 acres, the amount of land available. The total quantity of capital used must be less than or equal to $360, the amount available.

A further condition, important from the standpoint of the mathematics of programming but so obvious that a formal statement of it may seem unnecessary, is that no activity can be carried on at a negative level (produce a negative quantity of any of the three crops included in the planning model).

$$X_C \geq 0 \quad (2.4)$$
$$X_S \geq 0 \quad (2.5)$$
$$X_O \geq 0 \quad (2.6)$$

Equation 2.4 specifies that the units of corn produced must be greater than or equal to zero; i.e., nonnegative. Similar conditions for the quantities of soybeans and oats produced are specified in (2.5) and (2.6).

We next change the system of inequalities to one of equalities by adding disposal activities. These activities provide that any portion of the supply of any resource may go unused:

X_A = the quantity of unused land
X_L = the quantity of unused labor
X_M = the quantity of unused capital

Adding one disposal activity for each of the three resources—labor, capital, and land—to inequalities (2.1), (2.2), and (2.3) above, we arrive at the following set of equations:

$$1X_C + 1X_S + 1X_O + 1X_A + 0X_L + 0X_M = 12 \quad (2.7)$$
$$6X_C + 6X_S + 2X_O + 0X_A + 1X_L + 0X_M = 48 \quad (2.8)$$
$$36X_C + 24X_S + 18X_O + 0X_A + 0X_L + 1X_M = 360 \quad (2.9)$$

The production requirements or coefficients for the disposal activities may need interpretation. In (2.7), the digit 1 preceding X_A—the land disposal activity—means that permitting one acre of land to go unused subtracts one acre from the total available for other uses. However, not using one acre of land makes no demands on the labor or capital supply. Therefore, a zero precedes X_A in (2.8) and (2.9).

In programming we seek to find the values of X_C, X_S, X_O, X_A, X_L, and X_M that will maximize the sum of the products of these quantities and their prices. In other words, what combination of corn, soybeans, and oats should be produced and how much (if any) land, capital, and labor should go unused?

The problem then is to maximize C defined as return over variable cost where:

$$C = 40X_C + 30X_S + 20X_O \quad (2.10)$$

subject to the conditions imposed by (2.4), (2.5), and (2.6) and in (2.7), (2.8), and (2.9).

Students who have been exposed to matrix algebra will recognize its application to the solution of the problem. However, it is necessary only to be able to perform accurately the basic arithmetic operations to follow the steps in the simplex solution.

SIMPLEX SOLUTION TO CROP PRODUCTION PROBLEMS

The simplex method is a mathematical procedure (algorithm) that uses addition, subtraction, multiplication, and division in a particular sequential way to solve problems. The computations required in solving this very small model are lengthy; for a problem large enough to be realistic they are too burdensome to attempt with a desk calculator.

The orderly format in which the elements of the problem were arranged is the starting point for computations. All the basic data specified in the original problem (and presented earlier in Table 2.1) are given in Section 1 of Table 2.2. Sections 2 and 3 show how the computations proceed in the programming routine.

The columns and rows are as follows:

1. In the first section (at the start of the planning period) none of the resources is being used and no production is taking place. Thus all the resources (land, labor, and capital in this problem) are in disposal. The Net Price column which we will label "C" gives the net price of the disposal activities. As other activities enter the plan in subsequent sections, their net prices will take the place of the zero net prices originally assigned to the disposal activities.
2. The B column shows the level or amount of each resource at the outset. In later sections this column will indicate the level at which real activities as well as disposal activities enter the plan.
3. The activity columns indicate the amount of each of the resources required per unit of output. As before, the corn column specifies the amount of land required in the land row (1 acre in this case), the amount of labor required in the labor row (6 hours), and the amount of capital required in the capital row ($36).
4. The numbers in the disposal activity columns have the same meaning as those for the real activities. Thus to place one acre of land in disposal (nonuse) requires one acre of land, zero

TABLE 2.2: Simplex Linear Programming Problem

C (Net Price)	Resource or Activity	B (Resource or Activity Level)	Real Activities			Disposal Activities			R (Ratio)
			Corn	Soybeans	Oats	Land	Labor	Capital	
				SECTION 1					
0	Land	12	1	1	1	1	0	0	12
0	Labor	48	6	6	2	0	1	0	8
0	Capital	360	36	24	18	0	0	1	10
C (net price)			40	30	20	0	0	0	
Z (opportunity cost)			0	0	0	0	0	0	
Z-C (shadow price)		0	—40	—30	—20	0	0	0	
Value of program = $0									
				SECTION 2					
0	Land	4	0	0	2/3	1	—1/6	0	6
40	Corn	8	1	1	1/3	0	1/6	0	24
0	Capital	72	0	—12	6	0	—6	1	12
C (net price)			40	30	20	0	0	0	
Z (opportunity cost)			40	40	40/3	0	20/3	0	
Z-C (shadow price)		320	0	10	—20/3	0	20/3	0	
Value of program = $320									
				SECTION 3					
20	Oats	6	0	0	1	3/2	—1/4	0	
40	Corn	6	1	1	0	—1/2	1/4	0	
0	Capital	36	0	—12	0	—9	—9/2	1	
C (net price)			40	30	20	0	0	0	
Z (opportunity cost)			40	40	20	10	5	0	
Z-C (shadow price)		360	0	10	0	10	5	0	
Value of program = $360									

labor, and zero capital; hence the 1 in the land row and the zeros in the labor and capital rows. The meaning of the numbers in the R (ratio) column will be explained later.

5. The C row (net price) shows net price for each activity unit.
6. The Z row (opportunity cost) shows for each activity the value of other enterprises which must be sacrificed to produce one more unit of output. Since there are no real activities in the "Resource or Activity" column and all resources are going unused in the first section, adding one more unit to any activity would require no sacrifice of any of the others. Hence all entries in the opportunity cost row are zero.
7. The Z-C row (shadow price) indicates how adding an additional unit of any activity, including a disposal activity, will change the value of the program. The signs, however, are reversed. A minus $40 in the corn activity in the first section indicates that adding an acre of corn will increase the value of the program by $40.

The plan specified in Section 1 is feasible from the standpoint of the algebra involved. Producing nothing and letting all resources go unused results in all the variables being greater than or equal to zero when all the equations are satisfied. Of course income (the value of the program) is zero with this plan. Since we are seeking an optimum as well as a feasible plan, we begin a routine of substituting real activities for the disposal activities in the original plan.

In our example, this procedure leads quickly to the optimum plan. In an actual farm planning problem, the process is much more prolonged.

Programming Computations

1. The column with the largest negative Z-C value (addition to net income) is the *outgoing column* from the first section and the *incoming row* in the second section.
2. To determine the *level* at which the new activity (incoming row) will be brought into the second section, divide the resource quantities in the B column of the first section by the coefficients in the outgoing column. Each B quantity is divided by the coefficient in its row of the outgoing column and the quotient entered in the R column (ratio).
3. The smallest quotient so formed indicates the maximum level to which the incoming row in the second section can be increased.

4. The row (resource supply or activity) with the smallest R becomes the outgoing row. Thus the labor row in our example will not appear in the second section. In selecting the smallest R, do not consider negative numbers.
5. The next step is to compute the figures for the rows in Section 2, ignoring for the moment the entries for the C and R columns.
6. We designate an *outgoing pivot* in Section 1. This is the coefficient at the intersection of the outgoing column and the outgoing row. In this example it is the 6 at the intersection of the corn column and the labor row.
7. We compute the incoming row in Section 2 by dividing the entry in each column of the outgoing row (Section 1) by the outgoing row's pivot. Each quotient thus formed is entered in the corresponding column of the incoming row. Thus $48 \div 6 = 8$, the entry in the B column in the incoming row of Section 2, and $6 \div 6 = 1$ forms the quotient in the corn column of the same row, and so forth.
8. To compute any new row in the second section other than the incoming row, the C row, and the Z row, arrange the data in four columns labeled O, I, P, and N as follows:

O = coefficients in the corresponding row of old section
I = coefficients of the incoming row in the new section
P = coefficients at the intersection of the outgoing column and the row in the old section under consideration
N = coefficients for the "new row" of the new section.
These are computed from the formula: $O - (I \times P) = N$

In the example, the coefficients involved and the resulting new land row are as follows:

NEW LAND ROW, SECTION 2

O	I	P	N
12	8	1	4
1	1	1	0
1	1	1	0
1	$\frac{1}{3}$	1	$\frac{2}{3}$
1	0	1	1
0	$\frac{1}{6}$	1	$-\frac{1}{6}$
0	0	1	0

We follow the same pattern for the new capital row in Section 2. Note that the I column is identical to I in the preceding section:

NEW CAPITAL ROW, SECTION 2

O	I	P	N
360	8	36	72
36	1	36	0
24	1	36	–12
18	⅓	36	6
0	0	36	0
0	⅙	36	–6
1	0	36	1

The reader should note that the B column coefficient was included in the above routine; i.e., the top entries in both the O and I columns.

In this instance, the Z-C value which results in the B column is the value of the program. It may be computed routinely as below:

NEW Z-C ROW, SECTION 2

O	I	P	N
0	8	–40	320
–40	1	–40	0
–30	1	–40	10
–20	⅓	–40	–20/3
0	0	–40	0
0	⅙	–40	40/6
0	0	–40	0

9. Enter the appropriate prices in the C column and C row. Disposal activities will have zero prices and other activities the same price as entered in the C row at the start of the program. Thus land and capital in our example will have a zero price and corn a price of 40 in the C column.
10. The value of the program can be checked by multiplying each entry in the C column by the entry on the same row of the B column, Section 2. The sum of the products thus formed is the value of the program for Section 2, in this case $320:

$0 \times 4 = 0$
$40 \times 8 = 320$
$0 \times 72 = 0$
Total $320, the value of the program

11. We can find the optimum plan without computing the Z row, since we computed the Z-C value using the routine described above. However, you may find that computing the Z values provides you additional insight into the meaning of

the programming process. Additionally, the accuracy of the new Z-C values for each activity can be checked by comparing them with the difference between the Z and C values. Differences other than minor rounding errors indicate that an error or errors have occurred.

We can compute the Z row (opportunity cost row) for all columns of Section 2 (except B) following the same procedure used to calculate the value of the program. Multiply each entry in a column by the C column value in the corresponding row and sum the products. The entry in the Z row of the corn column in Section 2 becomes 40, obtained as follows:

$$\begin{array}{rcl} 0 \times 0 & = & 0 \\ 40 \times 1 & = & 40 \\ 0 \times 0 & = & \underline{0} \\ \text{Total} & & 40 \end{array}$$

The entry of 40 in the Z row of the soybean column in Section 2 comes from the following calculations:

$$\begin{array}{rcl} 0 \times 0 & = & 0 \\ 40 \times 1 & = & 40 \\ 0 \times -12 & = & \underline{0} \\ \text{Total} & & 40 \end{array}$$

12. When the plan fails to meet the test of no negative Z-C values, a new section must be formed. Exactly the same rules are followed as above except that the old section now will be Section 2 and the new one, Section 3. We continue the iterations until there are no negative Z-C values in the solution.

Interpretation of Final Section

The optimum solution in the example is reached at the completion of the third section. We know this is true because the net price of none of the three products exceeds its opportunity cost. No addition to net income, given the planning assumptions, can be achieved by any change from the pattern of 6 acres of oats and 6 acres of corn. The entry in the capital row of the B column tells us that $36 of capital went unused in the plan.

Note that soybeans did not enter the plan. As a matter of fact the Z-C row in our final section indicates that raising an acre of soybeans would reduce net income by $10. If for some reason we insisted on including an acre of soybeans in the plan, we would depress income the least by reducing corn by 1 acre.

In this substitution we would replace $40 net return from corn with a $30 net return from soybeans. However, if we attempt to substitute soybeans for oats, labor becomes restricting. Returning to the labor row in the original model (2.8) and ignoring disposal variables because no labor was in disposal we have:

$$6X_C + 6X_S + 2X_O = 48$$

If $X_C = 6$ and $X_S = 1$, then

$$\begin{aligned} (6)(6) + (6)(1) + 2X_O &= 48 \\ 2X_O &= 6 \\ X_O &= 3 \end{aligned}$$

Thus if soybeans were substituted for oats instead of corn, the plan would include 3 acres of oats, 1 acre of soybeans, and 6 acres of corn. With appropriate substitutions in the objective function (2.10), the reduction in net income is demonstrable. Thus,

$$\begin{aligned} C &= (40)(6) + (30)(1) + (20)(3) \\ C &= 330 \end{aligned}$$

The $330 is $20 less than the $350 which would have resulted from a soybean for corn substitution and $30 less than the $360 forthcoming from the optimum program.

Although the net price per acre for soybeans was greater than for oats, labor was a limiting resource; when returns from both scarce land and scarce labor were considered, the combination of corn and oats in this example was superior to the corn-soybean combination.

Also of interest are the entries in the Z-C row of the disposal activity columns in Section 3. The $10 entry under land indicates the effect on net income of letting an acre of land go unused. The $10 decrease can be explained by tracing out the changes which result in the basis variables when $X_A = 1$ instead of $X_A = 0$. See the discussion in "Algebraic Presentation of a Solution."

The $5 figure under labor also indicated the effect on net income from permitting one hour of labor to go unused. This $5 decrease can also be explained by tracing out the changes which result in the basis variables when $X_L = 1$ instead of $X_L = 0$.

Machine Computation

The model we have developed above is much too limited in both restraints and activities to have practical application. However, even with the limited number of enterprises and resources the computations required were substantial. Fortunately, computations can now be made rapidly and at low cost by electronic computers. Instead of the corn, oats, and soybean activities we used in our simple illustrations, a realistic application of the method would involve perhaps 25 or 30 cropping activities (including fertilization at different levels) and as many livestock activities. In place of the three restraints of land, labor, and capital we might have 3 or 4 types of land, 4 or 5 labor restrictions, several capital restrictions plus others arising from feed supplies, government programs, buildings and facilities, and management and risk considerations.

The difficult tasks in programming are defining meaningful restraints, developing accurate price expectations, and estimating realistic resource-to-product relationships. Once these judgments have been made and the data properly arranged, large complex models can be processed quickly and accurately.

ALGEBRAIC PRESENTATION OF SOLUTION

The computations in the simplex solution presented in the previous pages are based on relatively simple algebraic manipulations. In this section we follow through the same problem, giving greater emphasis to the algebraic interpretations of the simplex method.

We begin with the same four linear equations (2.7, 2.8, 2.9, and 2.10) we formed earlier in the chapter. The problem (i.e., restrictions, activities, coefficients, and prices) has not been changed in any way except that the objective function (2.10)—the equation containing the C or income variable—appears, for convenience, as the first equation in the set. Thus the first equation (2.11) in Step I corresponds to the C row or objective function in the example shown in Table 2.2.

Step I

$$C - 40X_C - 30X_S - 20X_O = 0 \quad (2.11)$$
$$1X_C + 1X_S + 1X_O + 1X_A = 12 \quad (2.12)$$
$$6X_C + 6X_S + 2X_O + 1X_L = 48 \quad (2.13)$$
$$36X_C + 24X_S + 18X_O + 1X_M = 360 \quad (2.14)$$

The four equations describe the restrictions we have imposed, the activities we wish to consider, and the coefficients we have specified in formulating the problem. Note that we are dealing with a system of four equations involving seven unknowns. There is a wide variety of values for the seven variables that would satisfy the equations. We are, however, looking for unique values which will maximize the C value (give the greatest income) given the objective function described by

$$C = 40X_C + 30X_S + 20X_O \tag{2.15}$$

With more unknowns than equations, we are confronted with a dilemma in our search for unique values. A simple example will serve to illustrate the problem. Following are two equations in three unknowns:

$$X_1 + 4X_2 = 12 \tag{2.16}$$
$$2X_2 + X_3 = 9 \tag{2.17}$$

Any of the following sets of values will satisfy the two equations: (a) $X_1 = 4$, $X_2 = 2$, and $X_3 = 5$; (b) $X_1 = 8$, $X_2 = 1$, and $X_3 = 7$; or (c) $X_1 = 5$, $X_2 = 1\frac{3}{4}$, and $X_3 = 5\frac{1}{2}$. There are many more sets which could serve equally well.

The problem of a greater number of unknowns than equations is a familiar one in linear programming analysis. Indeed, adding disposal variables insures that the number of equations is always less than the number of variables.

The simplex method provides a procedure whereby we can avoid a multitude of solutions in spite of the excess of variables over equations. The number of unknowns is reduced to the number of equations by assuming zero values for certain of the variables.

If we return to (2.16) and (2.17) where we have three variables, this technique can be demonstrated. We can assign values of zero to X_2 and record immediately that $X_1 = 12$ and $X_3 = 9$. If we elect to let $X_1 = 0$, we observe readily from (2.16) that $X_2 = 3$. Then we substitute for X_2 in (2.17).

$$\begin{aligned} (2)(3) + X_3 &= 9 \\ X_3 &= 9 - (2)(3) \\ X_3 &= 3 \end{aligned}$$

A process similar to the one described above is repeated many times in the simplex solution. However, we follow a set of rules which guides us in determining which unknowns will be assigned zero values. You are already acquainted with these rules from the solution of the programming problem in Table 2.2.

Returning now to the set of equations described by the initial section of Table 2.2 and restated above, the algebraic manipulation begins by assigning zero values to all real variables (activities). Thus X_C, X_S, and X_O have zero values in (2.11), (2.12), (2.13), and (2.14). Given this assumption we know immediately that $C = 0$, $X_A = 12$, $X_L = 48$, and $X_M = 360$. This is the initial plan where no production is undertaken, all resources are in disposal, and income is zero.

We now apply the same simplex rules used before to solve the problem. We inspect the objective function (2.11) to determine which variables (if any) not in the basic solution have a negative coefficient.[1] From these we select the coefficient with the largest negative value (—40). The variable to which this coefficient is attached promises the largest per unit gain in income (X_C in our example). Consequently, X_C is brought into the basis in Step II: i.e., it is assigned a nonzero value. Furthermore, it is assigned the largest value possible without violating any restraint.

To determine the latter value we take the ratios of the values of variables in the basic solution (in the first solution these values represent the quantities of resources available) to the coefficients for the entering variable. The coefficients in the first solution represent the resource requirement per unit of activity. The process described above is the same one we carried through in developing the R column in Section 1 of Table 2.2. From the ratios thus formed we select the minimum ratio, ignoring negative values. The minimum ratio occurs for variable X_L (labor); i.e., labor is the most restricting resource constraint. The value of X_L is zero in the upcoming or new solution; i.e., we use all of the available labor to produce 8 units of corn, X_C, which is the maximum we can produce with the labor available. The procedure described above is summarized in Table 2.3.

The next step is to divide (2.13) (the labor equation) by the number 6 (the coefficient for labor used in corn production) which provides the equation for the X_C variable in (2.20). At this stage, which we shall refer to as Step II, the system of equations is as follows [(2.11), (2.12), and (2.14) are presented below as (2.18), (2.19), and 2.21)]:

Step II

$$C - 40X_C - 30X_S - 20X_O = 0 \quad (2.18)$$

$$1X_C + 1X_S + 1X_O + 1X_A = 12 \quad (2.19)$$

$$1X_C + 1X_S + 1/3X_O + 1/6X_L = 8 \quad (2.20)$$

$$36X_C + 24X_S + 18X_O + 1X_M = 360 \quad (2.21)$$

1. The terms basis or basic solution are widely used in programming literature to describe that set of variables which have a nonzero value in a solution. At this stage only the disposal or slack variables X_A, X_L, and X_M are in the basis.

TABLE 2.3: Determining Most Restrictive Resource Restraint

Coefficient for Variable entering Basis* (X_C)	Value of Variables in Original Solution	R Value
1	$X_A = 12$	12
6	$X_L = 48$	8
36	$X_M = 360$	10

* No R value is computed from the first equation because it is the objective function or the criterion row.

You should observe that at this point we have assigned zero values to X_S, X_O, and X_L. We know that X_C has a value of 8 which is reflected in (2.20). We have yet to ascertain values for X_A, X_M, and C before we have a second plan. To do this we engage in algebraic manipulations similar to those described in finding values for the variables in (2.16) and (2.17).

We now rearrange (2.20) so that the value of X_C becomes explicit and substitute this value in each of the remaining three equations. In doing so we carry along all the terms in each equation even though they are zero valued. Following this procedure enables us to save information which will prove useful later.

Equation 2.20 with everything but X_C on the right-hand side becomes:

$$1X_C = 8 - 1X_S - 1/3X_O - 1/6X_L$$

Substituting this new X_C value in (2.18) gives us equation 2.22.

$$\begin{aligned} C - 40(8 - 1X_S - 1/3X_O - 1/6X_L) - 30X_S - 20X_O &= 0 \\ C - 320 + 40X_S + 40/3X_O + 40/6X_L - 30X_S - 20X_O &= 0 \\ C + 10X_S - 20/3X_O + 40/6X_L &= 320 \end{aligned} \quad (2.22)$$

Because X_S, X_O, and X_L have zero values, C = 320, the value of the program at this stage. Substituting the new X_C value in (2.19) provides equation 2.23.

$$\begin{aligned} (8 - 1X_S - 1/3X_O - 1/6X_L) + 1X_S + 1X_O + 1X_A &= 12 \\ 2/3X_O + 1X_A - 1/6X_L &= 4 \end{aligned} \quad (2.23)$$

Equation 2.20 is presented below as (2.24).

$$1X_C + 1X_S + 1/3X_O + 1/6X_L = 8 \quad (2.24)$$

Equation 2.25 results from (2.21) when the newly determined X_C value is substituted.

$$36(8-1X_S-1/3X_O-1/6X_L) + 24X_S + 18X_O + 1X_M = 360$$
$$288 - 36X_S - 36/3X_O - 36/6X_L + 24X_S + 18X_O + 1X_M = 360$$
$$-12X_S + 6X_O - 6X_L + 1X_M = 72 \quad (2.25)$$

Step III

$$C + 10X_S - 20/3X_O + 40/6X_L = 320 \quad (2.22)$$
$$2/3X_O + 1X_A - 1/6X_L = 4 \quad (2.23)$$
$$1X_C + 1X_S + 1/3X_O + 1/6X_L = 8 \quad (2.24)$$
$$-12X_S + 6X_O - 6X_L + X_M = 72 \quad (2.25)$$

We now have a second solution; i.e., we have either assumed or solved for unique values for all of the variables. X_S, X_O, and X_L have zero values. We assumed a zero value for X_S and X_O at the start and have not altered this assumption. X_L became zero (left the basis) when all labor in disposal was allocated to corn production. Thus, when we take account of these zero values we can observe readily from the equations above the values for the variables in the solution. They are:

$$C = 320$$
$$X_A = 4$$
$$X_C = 8$$
$$X_M = 72$$

We now apply the familiar simplex criterion to determine whether we can better this solution (i.e., increase the value of the objective function) described by (2.22). Since there is a coefficient in (2.22) with a negative value, we know that improvement in income is possible. We select the largest negative coefficient (in this case there is no choice since there is only one negative value) and proceed as before. More specifically, X_O will be brought into the basis. We take the ratios of the X_O coefficients to the values of the variables presently in the basis (from the R column) and select the smallest ratio. This indicates the variable (in this case X_A) which will leave the basis, thus joining that group of variables with a zero value. Our encounter with the R column also specifies the value that will be assigned to X_O as it enters the basis. A quick inspection of the coefficients for X_O in the Step III equations reveals that X_O will enter the basis with a value of 6. Variables with zero values at this stage are X_S, X_A, and X_L.

Now we proceed as before with the following results:

Step IV

$$C + 10X_S + 10X_A + 5X_L = 360 \quad (2.26)$$
$$X_O + 3/2X_A - 1/4X_L = 6 \quad (2.27)$$
$$1X_C + 1X_S - 1/2X_A + 1/4X_L = 6 \quad (2.28)$$
$$-12X_S - 9X_A - 9/2X_L + 1X_M = 36 \quad (2.29)$$

Equation 2.26 contains no negative values; hence we have reached an optimum solution. The value of the variables in the basis are as follows:

$$\begin{aligned} C &= 360 \\ X_O &= 6 \\ X_C &= 6 \\ X_M &= 36 \end{aligned}$$

In reading these values from the equations in Step IV, it is essential to remember that we gave zero values to X_S, X_A, and X_L at the start of the current iteration. Thus the solution obtained is exactly the same as that achieved in Section 3 of Table 2.2.

The coefficients of variables not in the solution also are of interest. The coefficients in (2.26) indicate the change in the value of the program which would result from assigning any variable not in the basis a value of 1. Assigning X_A a value of 1 gives this result:

$$\begin{aligned} C + 10 &= 360 \\ C &= 350 \end{aligned}$$

Thus, placing a unit of X_A in disposal would reduce income by $10.

The $10 decrease can be explained by tracing out the changes that result in the basis variables when $X_A = 1$ instead of $X_A = 0$. To facilitate the analysis we use (2.26), (2.27), (2.28), and (2.29). The equations including variables with zero coefficients are presented below to illustrate the similarity to coefficient arrangement of Section 3 of Table 2.2.

$$6 = 0X_C + 0X_S + 1X_O + 3/2X_A - 1/4X_L + 0X_M \quad (2.30)$$
$$6 = 1X_C + 1X_S + 0X_O - 1/2X_A + 1/4X_L + 0X_M \quad (2.31)$$
$$36 = 0X_C - 12X_S + 0X_O - 9X_A - 9/2X_L + 1X_M \quad (2.32)$$
$$C = 40X_C + 30X_S + 20X_O + 0X_A + 0X_L + 0X_M \quad (2.33)$$

When $X_A = 1$ is substituted in (2.30) in lieu of $X_A = 0$, and with terms of zero value eliminated, we have:

$$6 = X_O + (3/2)(1)$$

because $X_L = 0$ and the $-1/4X_L$ term $= 0$. Then

$$\begin{aligned} 6 - 3/2 &= X_O \\ 4\ 1/2 &= X_O \text{ (acres of oats in the revised plan)} \end{aligned}$$

From a similar substitution in (2.31) we obtain the effect of the land reduction on corn production:

$$6 = 1X_C - (1/2)(1)$$

where $X_S = 0$ and $X_L = 0$

$$6 + 1/2 = X_C$$
$$6\ 1/2 = X_C \text{ (acres of corn in the revised plan)}$$

To estimate the impact on capital in disposal we use (2.32):

$$36 = (-9)(1) + X_M$$

where $X_S = 0$ and $X_L = 0$

$$36 + 9 = X_M$$
$$45 = X_M \text{ (amount of capital in disposal in the revised plan)}$$

To adjust to the one-acre reduction in land we have reduced oats from 6 to 4½ acres and increased corn from 6 to 6½ acres. To verify that the reduction in the value of the program would be the $10 indicated in the Z-C row we substitute the revised values for the basis variables in (2.33).

$$C = (40)(6\ 1/2) + (30)(0) + (20)(4\ 1/2) + (0)(45)$$
$$C = 260 + 90$$
$$C = 350$$

In following the path indicated by the coefficients in the final section we have achieved an optimum adjustment to the decrease from 12 to 11 acres in the amount of land in use. Meeting the reduction by any other combination of activities would result in a greater reduction in the value of the program.

We noted earlier that not all the $360 capital available at the start has been used. Thus capital has not limited the plan. Making more capital available (with no additional land or labor) would add nothing to income, as indicated by the zero in the Z-C row of the capital disposal activity.

Thus the programming procedure not only indicates the most profitable plan but supplies estimates of the marginal return of each resource or restraint included in the model. The interpretation of the shadow prices on the disposal activities as marginal value productivities of the resource restraints is not consistent with a rigorous definition of the marginal product. Strictly speaking, the marginal physical product of an input is that increase in total output of an activity associated with an increment of the input with all other inputs held constant. The latter condition cannot be met within the linear programming framework, because the production coefficients for the activities are defined in fixed ratio to one another. Thus the use of one input cannot be increased without an increase in another. In spite of this logical weakness, the concept of the shadow price as an estimate of the marginal value product is operationally useful within the context of farm planning models. We will continue to equate the two concepts in future references. This is important because it provides information to the farm manager on what resources he can best expand to increase income. For example, the programming process might indicate that he could afford to pay very high wages to attract a hired man to break the spring bottleneck on labor.

MINIMUM RESTRAINTS

The restraints encountered thus far have been maximum restraints. Resource restraints are characteristically of this type. They specify that the activities can utilize a given quantity of a resource but no more. For example, a labor restraint typically would impose the following condition:

$$48 \geq 6X_1 + 6X_2 + 2X_3$$

where X_1, X_2, and X_3 represent activity levels and the coefficients indicate the quantity of labor required per unit of activity. This restraint does not specify that all labor must be utilized. By adding a disposal activity, X_4, which permits nonuse of labor, we form an equation of a type we have already encountered:

$$48 = 6X_1 + 6X_2 + 2X_3 + 1X_4 \tag{2.34}$$

The reader should note that the disposal variable in this equation has a positive sign. This indicates that some *positive* quantity of labor not required by activities X_1, X_2, or X_3 may go unused in X_4. The restriction as formed permits *underuse* of labor, and the value of X_4 indicates the extent of its underutilization.

But restrictions may be formed to permit *overfulfillment* of the restriction but not underfulfillment. A government program may require a minimum acreage of soil-conserving crops but *permit* a greater acreage. This condition might be expressed algebraically as follows:

$$30 \leq 2X_1 + 3X_2 + 4X_3 \qquad (2.35)$$

where 30 is the minimum acreage of forage required and X_1, X_2, and X_3 represent activities whereby this requirement can be met. Once again we must form an equation from the inequality before the simplex programming routine can begin. However, in this case we wish to permit overfulfillment of the restriction but not underfulfillment. Thus, as we form the disposal activity, we give it a negative sign:

$$30 = 2X_1 + 3X_2 + 4X_3 - 1X_4 \qquad (2.36)$$

Now any forage produced in excess of 30 acres in X_1, X_2, and X_3 can be offset by the slack permitted by the negatively signed disposal activity and the equation preserved. On the other hand, any deficit of forage produced in real activities X_1, X_2, and/or X_3 cannot be compensated for in a positive X_4 (disposal variable) value.

The use of minimum restrictions described above give rise to an added complexity in the simplex computing routine. To illustrate this problem we return to the original crop production model. The initial section of this problem was as follows:

	B	P01	P02	P03	R01	R02	R03
C		40	30	20	0	0	0
R01	12	1	1	1	1	0	0
R02	48	6	6	2	0	1	0
R03	360	36	24	18	0	0	1

As we indicated in previous discussion of the simplex method, a feasible solution is specified in the original formulation of the problem. All resources are unused and each of the real activities is carried on at zero level. The initial solution of course is not an optimum solution because the value of the program is zero. However, it is the starting point from which the simplex method proceeds to find a better and eventually an optimum solution.

Suppose we add a minimum restraint requiring that at least 2 acres of soybeans be included in the program. Thus the initial tableau would be as follows:

	B	P01	P02	P03	R01	R02	R03	R04
C		40	30	20	0	0	0	0
R01	12	1	1	1	1			
R02	48	6	6	2		1		
R03	360	36	24	18			1	
R04	2	0	1	0	0	0	0	−1

At this point we have no readily feasible solution. The conditions specified for equations R01, R02, and R03 could be met as before by placing the entire quantity of each resource in disposal (X_1, X_2, $X_3 = 0$ and $X_4 = 12$, $X_5 = 48$, $X_6 = 360$).

These three equations are as follows:

$$12 = X_1 + X_2 + X_3 + X_4 \quad (2.37)$$
$$48 = 6X_1 + 6X_2 + 2X_3 + X_5 \quad (2.38)$$
$$360 = 36X_1 + 24X_2 + 18X_3 + X_6 \quad (2.39)$$

The fourth row (R04) offers difficulty because it provides for meeting the equation specified only by resorting to a negative quantity of soybean production (less than zero quantity of soybean production capacity would be utilized) in the "initial solution."

$$2 = 0X_1 + 1X_2 + 0X_3 - X_7 \quad (2.40)$$

Because X_1, X_2, and X_3 by assumption have zero values in the original matrix,

$$-X_7 = 2, \text{ or } X_7 = -2 \quad (2.41)$$

This of course makes no sense. The programming solution protects against a result that would specify using less than zero acres of soybean production capacity or using less than zero quantity of labor by specifying that all variables entering the solution must be non-negative. Thus a solution which requires a negative value for one of the real or disposal activities is not a feasible solution.

Because we need a basis from which to begin the simplex routine, we must adjust our original formulation of the model. We do this by providing still another activity which we call an artificial activity. The reader may prefer to look upon it as a second disposal activity. We bypass R04 temporarily by assigning it a zero value (as we have P01, P02, and P03) in the original basis. The artificial activity, however, is one of the variables in the original solution. To make sure that the artificial activity does not appear in the final solution, we attach a large negative price to it. Then we proceed as before with the simplex routine. The first iteration will involve solutions with a negative "value of the program" because of the large penalty attached to the artificial activity.

	B	P01	P02	P03	R01	R02	R03	R04	A04
C		40	30	20	0	0	0	0	—1000
R01	12	1	1	1	1				
R02	48	6	6	2		1			
R03	360	36	24	18			1		
A04	2	0	1	0	0	0	0	—1	1

This large negative price will also result in the artificial activity dropping out of the basis early in the computing process. Often models with one or more artificial activities may require several more iterations to arrive at an optimum solution than one would expect if all restrictions were maximum restraints.

Most linear programming computer routines automatically enter an artificial activity when a minimum restriction is specified in the model. In the solution of such models the routine may sometimes go through several iterations before a positive value of program appears. Due to these added iterations, minimum restraints are typically more costly in terms of computing time than maximum restraints.

EQUALITY RESTRAINTS

Thus far we have talked about maximum (less than or equal to) restraints and minimum (greater than or equal to) restraints. There is still a third type—equality restraints—which permit neither underfulfillment nor overfulfillment of the restraint. Thus no disposal or slack activities attach to equality restraints in the model.

Although equalities are imposed less frequently than maximum or minimum restraints in farm planning models, where appropriate they are useful and relatively simple to apply. Government programs which require diverting an exact number of acres give rise to one frequent application. Assume you must divert exactly 70 acres and that X_D constitutes a land-diverting activity; then:

$$70 = 1X_D \tag{2.42}$$

(where the unit of activity X_D is 1 acre) becomes the equation built into the model. In the restraint row the B column coefficient is 70 and the coefficient for the X_D activity $+1$. There is no slack variable.

To cite another example, a farm operator may wish to include in the cropping program exactly 300 acres of row crops, no more and no less. Let $X_1 = 1$ acre of continuous corn production, $X_2 =$ a corn-corn-oats-meadow sequence where 4 acres equals one unit of activity, and $X_3 = 1$ acre of continuous soybean production; then the equality

$300 = X_1 + 4X_2 + X_3$ is formed to restrain row crops to the specified acreage. Again no slack variable is included.

The solution of models containing equality restraints requires resort to artificial activities, formed exactly like those for minimum restraints. The penalty price attached to the artificial activity assures its leaving the basis early in the computation process after serving its function as a starting point in the search for the optimum solution.

DUAL SOLUTION

It is possible to obtain the same information from a programming routine through quite a different approach called the dual solution. This approach emphasizes the relationship that exists between the prices of the products and the marginal value products of the resources. The sum of the marginal values of the resources used to produce a product must equal the price of the product. The reader will recall that only those resources that are in short supply are imputed a value greater than zero. Those resources (restraints) which appear in the solution–i.e., some part of which is placed in disposal–are imputed a zero value in the closed system that characterizes a programming model. The first step in the dual is to formulate the problem to ensure that the value of the resources used to produce one unit of the product is at least as great as the price of the resources. The objective of the programming routine then becomes one of selecting the set of values for the mix of inputs that minimizes the total "value" imputed to all resources used to produce the entire product mix.

We will illustrate the dual approach by organizing within the dual framework the crop selection problem presented at the beginning of this chapter. First we form a set of inequalities which state that the price of the product must be less than or equal to the sum of the products formed when the production inputs are multiplied by their marginal values:

$$40 \leq 1X_1 + 6X_2 + 36X_3$$
$$30 \leq 1X_1 + 6X_2 + 24X_3$$
$$20 \leq 1X_1 + 2X_2 + 18X_3$$

The relationships shown above may be altered as before to form equations by the addition of disposal activities. However, in this case the coefficients for the disposal activities provide for the amount by which the prices fall short of the value of resources they require. Thus they have negative signs and the system of equations becomes:

TABLE 2.4: Dual Formulation of Crop Problem, Original Tableau

Activity	B (Net Price)	Resources			Disposals			Artificials		
		Land	Labor	Capital	Corn	Soybeans	Oats	Corn A	Soybeans A	Oats A
Corn (A)	40	1	6	36	—1	0	0	1	0	0
Soybeans (A)	30	1	6	24	0	—1	0	0	1	0
Oats (A)	20	1	2	18	0	0	—1	0	0	1
C (resources)		+12	+48	+360	0	0	0	+M	+M	+M
Z (opportunity cost)		0	0	0	0	0	0	0	0	0
Z-C (shadow price)		—12	—48	—360	0	0	0	—M	—M	—M

$$40 = 1X_1 + 6X_2 + 36X_3 - 1X_4$$
$$30 = 1X_1 + 6X_2 + 24X_3 - 1X_5$$
$$20 = 1X_1 + 2X_2 + 18X_3 - 1X_6$$

At this stage we pause to compare the dual model with the previous formulation of the problem (see Table 2.4). The original tableau of the dual linear programming problem is presented in Table 2.2. We note the following differences:

(1) The B column entries are the prices of the crops instead of resource quantities.

(2) The resources used in production, not the cropping alternatives, are the activities.

(3) The $X_1 \ldots X_n$ values which we seek to determine in the programming process represent values imputed to the resources, not quantities of crops.

The dual formulation of the problem differs in still another important way from the previous model. The quantities of resources available constitute the C row. Previously we were attempting to find the combination of crop activities that would maximize the value of the program. Now we seek the relative values that can be imputed to the resources (land, labor, and capital) which will minimize the total value imputed to the resources used in production. The latter quantity is the value of the program. It is equal in this example to the total quantity of land times the value imputed to land plus the quantity of labor times the value imputed to labor plus the quantity of capital times the value imputed to capital.

The model thus formulated can be solved with the simplex routine. However, since we are interested in minimizing the total value imputed to resources, we select the highest positive Z-C value as the outgoing column and the incoming row in the new section. The computation process is complete when there are no longer any remaining positive Z-C values. Also, because we have minimum restrictions in the model, a further modification—the addition of artificial activities—is necessary. Because the formulation of artificial activities has been discussed in a previous section, the problems they entail will not be recounted here.

The solution to the dual model indicates the value imputed to each resource. The value of the program is the total value of resources used—the quantity we sought to minimize. In the cropping problem our solution indicates that land has a marginal product of 10, labor 5, and capital 0.

The Z-C values attached to the product now indicate the amount of crop that will be produced and the amount of resources, if any, in disposal—in this case 6 acres of corn, 6 acres of oats, and $36 in disposal.

The total value of crops produced equals 6 × $40 (acres of corn and net price of corn) plus 6 × $20 (acres of oats and net price of oats). Likewise the quantity of land times its imputed price (12 acres × $10) plus the quantity of labor times its imputed price (48 hours × $5) equals $360. (Capital, because some goes unused, has an imputed price of zero.) Thus the programming process assigns marginal values to resources such that the value of production is just exhausted when imputed to inputs used in the production process.

CHAPTER THREE

CONSTRUCTING PROGRAMMING MODELS

You have probably already sensed that a planning model consists of highly interdependent components. Indeed a programming model is such a highly integrated system that it is difficult to discuss one component without involving others. Thus, although this chapter is divided into separate sections dealing with activities, restraints, coefficients, and the objective functions, all four components of necessity frequently are discussed out of sequence.

DEFINING ACTIVITIES

We have already encountered the three major categories of activities: (1) real, (2) disposal, and (3) artificial. Once you understand the role of disposal and artificial activities and are familiar with how and why they are used, there is little point in concerning yourself further with them. Most modern computing routines contain provisions for supplying the needed disposal activities. The programmer in preparing his computer input need only distinguish among maximum, minimum, and equality restrictions. Likewise, most modern computing routines automatically provide for the addition of artificial activities where they are needed, with no initiative required on the part of the model builder. In this chapter we include disposal and artificial activities in the partial models which introduce minimum and equality restraints. This is done only to remind you that they do differ from the more familiar maximum restraints.

Real activities take a variety of forms. Among the most common are the following:

1. Producing or growing crops
2. Raising and/or feeding livestock
3. Selling products
4. Buying or hiring inputs or services including labor and capital
5. Harvesting crops

6. Transferring inputs or intermediate products from one activity or time period to another.
7. Paying fixed costs and/or family living expenses

One can also combine several functions within a single activity. Perhaps the most frequent combination is one of growing, harvesting, and marketing. Confusion often results when the range of functions being combined into an activity is not sharply defined. For this reason it is well to analyze carefully the range of functions you wish to include within each activity as you are building the model.

The activity unit—the amount of crop or livestock production, labor hiring, or capital borrowing each unit of activity represents—should also be clearly defined at the beginning of the model building process. The planner is accorded great flexibility in the units he selects. The critical factor is not the definition given the activity unit but the consistency with which the definition is adhered to in forming all coefficients for the activity. One can save time and energy in model building if he defines activity units with this in mind. For example, rather than specify one bushel of corn as the activity unit for a corn growing and harvesting activity, it is more convenient to use one acre. When developing a crop activity that encompasses a cropping sequence such as CCOMM, the activity unit conveniently can encompass five acres—two acres of corn, one acre of oats, and two acres of meadow. In the case of swine production, 100 pounds of pork, a finished hog, or one or more litters of finished hogs may comprise the activity unit. However, defining coefficients typically is facilitated by focusing on one litter (or in the case of a two-litter system on two litters) of finished hogs.

The characteristic that distinguishes one activity from another is not the name we attach to it but the production and other coefficients that appear in the rows of the model. To be separate and distinct, two activities must have different coefficients in one or more rows. Where we wish to compare the profitability of alternative methods of producing the same product, we form separate activities—one for each method being compared. Likewise, we can compare alternative production schedules, levels of fertilizer, or rates of planting by defining each schedule, level, or rate as a separate activity.

RESTRAINTS

Restraints may also be referred to as constraints or restrictions.

All three terms have essentially the same meaning. There are three basic types of restraints: (1) maximum, (2) minimum, and

(3) equality. The characteristics of the three types were discussed in Chapter 2.

Restrictions also may be classified on the basis of their purpose. Who or what gives rise to the restriction? The major classes are as follows:

1. Resource or input restrictions: included are such items as land, capital, labor, and facilities for livestock production.
2. External restrictions: this category encompasses such items as acreage allotments in government programs and limits on credit imposed by lenders.
3. Subjective restrictions: these restraints are imposed by the operator himself. Their limits may be difficult to define, but they frequently are both real and significant to the planning process. Often the restraints imposed stem from basic business and personal objectives of the operator. Among restraints of this type are the following:
   a. Limitations on the amount of credit the operator is willing to use. In many cases this is less than the quantity his creditors are willing to extend. The typical motivation for such a limitation is a vague and poorly defined desire to avoid the hazards of debt.
   b. Risk restraints on the level of livestock activities with a record of highly variable income patterns such as feeding lambs and heavy cattle.
   c. Minimum restraints on activities the operator considers desirable for nonincome reasons such as purebred beef cows, dairy cows, or soil-conserving crops.

TRANSFER ROWS

Transfer equations, as the label suggests, occupy rows in the model in a manner similar to restrictions, but the similarity ends there. Transfer rows provide a vehicle whereby the services or output of one activity may be transferred in the model to another activity. Thus one may transfer corn from a corn growing and harvesting activity to a hog producing and selling activity. Labor may be transferred from a labor hiring activity to any activity requiring labor. The typical B column entry for a transfer row is zero. However, if the model builder wishes, he can provide inventories of feed and supply on hand through the B column. Where this is done the B column entry will be nonzero.

COEFFICIENTS

The coefficients specify how the magnitude of a restraint (or transfer row) will be influenced by an increase of one unit of each activity in the model. In the simplest and most common case the coefficient reflects the demand one unit of activity makes on the resource represented by the row in which the coefficient appears. By convention, coefficients signifying a demand on a row carry a positive sign.

However, certain coefficients may signify that an activity will "add to" the supply of resource (or other restricting elements) represented by a row. The coefficients at the intersection of a grain buying column and a grain transfer row or a labor hiring column and a labor row are examples. These "add to" coefficients should carry a negative sign in the row or rows they are augmenting.

Confusion concerning signs is one of the most frequent problems encountered among beginning programmers. It is helpful to look at each row of a model as an equation.

B	X_1	X_2
200	3	—1

Let B represent the left-hand side of the equation and X_1 and X_2 variables the right side. From the partial model above $200 = 3X_1 - X_2$ where:

X_1 = a production activity which uses resource B at the rate of 3 units of resource per unit of activity.
X_2 = an activity, perhaps a buying activity, which adds to the supply of B.

If X_1 enters at (say) a level of 2, it will subtract (3) (2) = 6 units of resources from B since its sign will become a minus as it is shifted from the right to the left side of the equation.

$$200 = 6 - X_2$$
$$200 - 6 = -X_2$$

Furthermore, if X_2 enters at (say) a level of 100, it will add (—1) (100) = —100 to the supply of B since its sign will become a plus as it is shifted from the right to the left side of the equation.

$$200 = 3X_1 - 100$$
$$200 + 100 = 3X_1$$

What we have said above concerning signs does not apply straightforwardly to the coefficients entered in the objective function or C row. The problem of assigning appropriate signs to the objective function will be discussed in the following section.

THE OBJECTIVE FUNCTION

We shall refer in this section only to models where the objective is the *maximization* of some income variables which may range from net to gross income, depending on how the model is structured. Programming can also be applied to cost minimization problems and hence have quite a different set of criteria for optimization. But this is another matter, and in the interest of clarity we have postponed the introduction of minimization concepts until later.

The entries in the C row (i.e., the coefficients of the objective function) indicate how the total value of the solution will be altered by the addition of one unit of activity in the initial phase of the iterative process leading to the simplex solution. In its simplest form it represents the expected price of one unit of activity. Thus in a corn selling activity where the unit is one bushel, the C row value is simply the anticipated price of a bushel of corn. The sign is positive. A corn buying activity would carry a negative sign, since buying corn per se will reduce the value of the program.

Arriving at a net price for an activity encompassing several functions such as raising and selling corn is more complex. Typically the unit of activity will be acres. Thus one must first determine the expected yield and multiply this quantity times the expected price. The product is the estimated gross income. From this must be subtracted the cost of all inputs which (1) will not be drawn from the original B column quantities or (2) will not be debited through a system of purchasing activities and appropriate transfer rows in the model.

Proper interpretation of the objective function in the final solution is also a function of the type of model we structure. If we make provisions for meeting nonallocable costs in the model itself, then the value of the program (the magnitude of the final solution) represents a return to be imputed to those resources represented by the B column entries in the original model. However, if no provisions are made for "fixed cost paying" activities in the model, then fixed cost must be subtracted from the program value before we have an estimate of income to be imputed to the B column resources.

LENGTH OF PLANNING PERIOD AND CAPITAL RESTRAINTS

Discussion of two troublesome and interrelated problems—the specification of the planning period and the forming of capital restraints and coefficients—is postponed intentionally at this point. These two aspects of programming cause difficulty even among experienced programmers. The reader will be in a better position to cope with them after he has become thoroughly familiar with the concepts illustrated in the rest of this chapter. At this stage he should assume he is concerned only with an annual production cycle and capital coefficients which represent either the average annual capital requirement of an activity or its requirement at some predetermined peak point in its production cycle.

MODEL 3.1: *Combined Producing, Harvesting, and Selling Activity*

Explanation

The programmer has great flexibility in the manner in which he defines activities. It is important to define the activity precisely and then to form coefficients, including the net price, that are consistent with the definition.

Model 3.1 illustrates an activity that combines corn raising, harvesting, and selling. We use only two restrictions in the model—land and labor—for the sake of simplicity.

P01 = continuous corn raising, harvesting, and selling activity. Each unit of activity equals one acre.
R01 = land suitable for continuous corn production. The B column unit is acres.
R02 = a labor restraint. The B column unit is hours.

The row type should be designated for each row in the model. The objective function or C row is always labeled with an N, maximum restraints with an L, minimum restraints with a G, and equality restraints with an E.

			Real Activity
Row Type		B	P01
N	C		76
L	R01	320	1
L	R02	2,400	5

Points to Observe

1. In a realistic planning model there would be additional activities and restraints. They have been omitted to permit you to focus on the problem being illustrated. The disposal portion of the matrix, although formed in a straightforward manner, is omitted because most machine programs do not require its specification once the row type has been properly indicated.
2. In using machine computation, it is likely that the routine with which you are working will require only that you designate each restraint as either a maximum, minimum, or equality restraint. The rows have been labeled to indicate their type. N designates the row as the objective function and L signifies a maximum restraint.
3. In forming the C row value the expected yield of corn, 90 bushels per acre, has been multiplied by the expected price, $1.20 per bushel. This product, $108, represents the gross price of each unit of activity. The allocable cost of production not otherwise accounted for in the model has been subtracted to form the net price. In this case labor and land costs are not subtracted because the coefficient in each row represents a debit against the initial land and labor supply in the B column. However, $32 of other costs such as fertilizer, tractor fuel, seed, and herbicides have been deducted. Hence the net price is $108 — $32 = $76.
4. No provisions have been made for subtracting fixed costs. After fixed costs have been deducted, the value of the program (the maximum value of the objective function in the solution) represents a net return to the land and labor quantities specified in the original B column.

MODEL 3.2: *Selling Activity*

Explanation

One activity described in this model encompasses corn growing and harvesting. The output from this activity is placed in a transfer row. A corn selling activity has been included which provides for selling the corn. This formulation has no advantage over a combined growing, harvesting, and selling activity where the structure of the model is as simple as illustrated. However, it introduces the concept of the transfer row which will have great utility later as we attempt to build more realistic models.

P01 = a corn growing and harvesting activity. The unit of activity is one acre.
P02 = a corn selling activity. The unit of activity is one bushel.
R01 = a land restraint. The B column unit is acres of land.
R02 = a labor restraint. The B column unit is hours of labor.
R03 = a corn transfer row. The transfer unit is bushels.

Row Type		B	P01	P02
N	C		—32	1.20
L	R01	320	1	
L	R02	2,400	5	
L	R03	0	—90	1

Points to Observe

1. The P01 coefficient in the transfer row (R03) requires a negative sign. The 90 represents the bushels of corn harvested per unit of activity. It is negative because it contributes to the supply in R03. Note that it would be plus if transposed to the B side of the equation.
2. The P02 coefficient in R03 carries a plus sign. Selling corn takes away from the supply in the transfer row. If transposed to the B side of the equation, it would become negative.
3. The net price shown in the C row for the corn growing and harvesting activity is negative. The latter functions, when considered separately, subtract from income. The income from corn derives from its sale through P02 whose coefficient in the C row is positive. Thus the C row entry for P01 is negative by the amount of allocable costs, other than land and labor, entailed in corn growing and harvesting.
4. The P02 value in the C row is positive. Corn selling adds to the value of the program. The magnitude of the P02 value is the expected price of corn per bushel minus any costs incurred in marketing.

MODEL 3.3: *Activity Using an Intermediate Product*

Explanation

This model is identical to Model 3.2 except that a hog raising and selling activity has been added. This permits a choice between selling or feeding the intermediate product, corn, in the process of obtaining a solution.

P01 = a corn growing and harvesting activity. The unit of activity is one acre.
P02 = a corn selling activity. The unit of activity is one bushel.
P03 = a hog raising and selling activity with one farrowing per year. The unit of activity is one sow and a litter of seven pigs.
R01 = a land restriction. The B column unit is acres.
R02 = a labor restriction. The B column unit is hours.
R03 = a corn transfer row. The transfer unit is bushels.

Row Type		B	P01	P02	P03
N	C		−32	1.20	205.20
L	R01	320	1		
L	R02	2,400	5		16
L	R03	0	−90	1	104

Points to Observe

1. The only new activity introduced is hog raising and selling. The P03 coefficient in the corn transfer row (R03) is positive since the hog raising activity requires corn.
2. The net price for P03 is positive. It is estimated as follows:

Seven 220-pound hogs at 18 cents per pound	$271.20
Increase in value of the sow	4.00
	$275.20
Less variable cost of 7-pig litter	70.00
	$205.20

3. The cost of corn consumed by the litter in P03 is not subtracted from the gross value in estimating net price. The 104 bushels of corn required per unit of activity is accounted for by the P03 coefficient in the R03 (corn transfer) row.

MODEL 3.4: *Alternative Harvesting Methods*

Explanation

This model illustrates a structure which permits corn to be harvested either as silage or as grain. A similar design would permit a choice among harvesting methods where other crops are involved.

P01 = a corn raising activity. The unit of activity is one acre.
P02 = a harvesting-corn-for-grain activity. The unit of activity is one acre.
P03 = a harvesting-corn-for-silage activity. The unit of activity is one acre.
R01 = a land restriction. The B column unit is acres.

R02 = a labor restriction. The B column unit is hours.
R03 = a standing (unharvested) corn transfer row. The transfer unit is acres.
R04 = a corn grain transfer row. The transfer unit is bushels.
R05 = a corn silage transfer row. The transfer unit is tons.

			Real Activities		
Row Type		B	P01	P02	P03
N	C		—25	—7	—18
L	R01	320	1		
L	R02	2,400	4	1	4
L	R03	0	—1	1	1
L	R04	0		—90	
L	R05	0			—14

Points to Observe

1. The model is obviously incomplete as it stands since no provision is made for utilization of either grain or silage. Activities which utilize corn and silage have been omitted to permit the illustration to focus on the key points.
2. The concept of transferring a crop prior to harvest has been introduced for the first time. P01 provides only for growing the corn. R03 transfers unharvested corn to the grain and silage harvesting activities.
3. The transfer unit in R03 is one acre of unharvested corn. The unit in R04 is bushels and in R05, tons.
4. The unit of activity for both P02 and P03 is one acre. The output from one unit of the grain harvesting activity (P02) is 90 bushels which is entered in R04, the corn grain transfer row. The output from the silage harvesting activity (P03) is 14 tons which is entered in R05, the silage transfer row.
5. The C row value in column P01 reflects only the allocable costs of corn production until harvest. The C row entries in P02 and P03 reflect respectively the cost of harvesting corn as grain and as silage.
6. The corn grain and corn silage harvesting activities require an acre of standing corn.

MODEL 3.5: *Crop Rotation Activity*

Explanation

Where there is interaction among crops such that the yield of one crop is a function of the rotation within which it is grown, frequently it is desirable to define crop activities in terms of a rotation or sequence. Because corn in a CCOM sequence is likely to yield slightly more and respond differently to fertilizer than does continuous corn, it is well to define the entire CCOM sequence as a single activity. The assumption is made that, should the activity enter the plan, first-year corn, second-year corn, oats, and meadow would be raised each year in equal proportions.

P01 = a CCOM activity. The functions involved are raising and harvesting corn and oats and raising meadow. The unit of activity is four acres.
R01 = a land restriction. The B column unit is acres.
R02 = a labor restriction. The B column unit is hours.
R03 = a corn transfer row. The transfer unit is bushels.
R04 = an oats transfer row. The transfer unit is bushels.
R05 = a meadow transfer row. The transfer unit is acres.

			Real Activity
Row Type		B	P01
N	C		—88
L	R01	320	4
L	R02	2,400	23
L	R03	0	—190
L	R04	0	—60
L	R05		—1

Points to Observe

1. The P01 coefficient for land (R01) is 4 because the unit of activity for P01 is four acres: one acre each of first-year corn, second-year corn, oats, and meadow. The P01 coefficient for labor (R02) likewise reflects the labor required to produce all four acres of the crops in the sequence.
2. The P01 coefficient in R03 (corn transfer) reflects the production from two acres of corn.
3. The P01 coefficient in R04 (oats transfer) represents the production of one acre of oats. Because meadow is transferred as an unharvested crop, the P01 coefficient in R05 is one acre.

MODEL 3.6: *Combination Crop Rotation and Selling Activity*

Explanation

In this case we consider a corn-corn-soybean cropping sequence. The activity is defined as corn growing and harvesting and soybean growing, harvesting, and selling. Note that the activity encompasses growing and harvesting corn but growing, harvesting, and *selling* soybeans.

P01 = a corn growing and harvesting and a soybean growing, harvesting, and selling activity. The unit of activity is three acres, consisting of two acres of corn and one acre of soybeans.
R01 = a land restriction. The B column unit is acres.
R02 = a labor restriction. The B column unit is hours.
R03 = a corn transfer row. The transfer unit is bushels.

			Real Activity
Row			
Type		B	P01
N	C		–6
L	R01	320	3
L	R02	2,400	14
L	R03	0	–178

Points to Observe

1. The output from corn growing is transferred to other activities through R03, but activity P01 includes selling soybeans.
2. The net price in the C row is obtained by estimating the gross income from soybean selling and deducting from it the variable costs not accounted for in the model for all components of the activity. In this example the gross income from soybean production is 30 bushels × \$2.60 per bushel = \$78. The variable costs are \$32 for first-year corn, \$34 for second-year corn, and \$18 for soybeans. Thus the total variable cost is \$84 per unit of activity, i.e. per three acres. Subtracting this from \$78 gives a net price of –\$6 per activity unit.

MODEL 3.7: *Multiple Land Restraints*

Explanation

Frequently the land within the farming unit varies considerably in its characteristics and production capabilities. Where land is clearly not homogeneous, it should be classified and each class treated as a separate land restriction. Corn or any other crop raised on Class I land is defined as an activity distinct from the same crop raised on Class II land. Corn on Class I land typically will have a larger coefficient in the corn transfer row than will corn raised on Class II land. Other coefficients—including labor, capital, and the C row variable cost—may differ. Furthermore, it is often desirable to provide for a different range of cropping activities on each class of land.

P01 = a continuous corn growing and harvesting activity on Class I land. The unit of activity is one acre.
P02 = a corn-corn-soybean growing and harvesting activity on Class I land. The unit of activity is three acres.
P03 = a CCOM activity on Class II land. The unit of activity is four acres. The activity includes growing and harvesting corn and oats and growing meadow.
P04 = a CCOMM activity on Class II land. The unit of activity is five acres. The activity includes growing and harvesting corn and oats and growing meadow.
R01 = a Class I land restraint. The B column unit is acres.
R02 = a Class II land restraint. The B column unit is acres.
R03 = a labor restraint. The B column unit is hours.
R04 = a corn transfer row. The transfer unit is bushels.
R05 = an oats transfer row. The transfer unit is bushels.
R06 = a soybean transfer row. The transfer unit is bushels.
R07 = a standing meadow transfer row. The transfer unit is acres.

			Real Activities			
Row Type		B	P01	P02	P03	P04
N	C		—32	—84	—82	—86
L	R01	240	1	3		
L	R02	80			4	5
L	R03	2,400	5	14	16	19
L	R04	0	—90	—178	—185	—190
L	R05	0			—60	—60
L	R06	0		—30		
L	R07				—1	—2

Points to Observe

1. The land coefficients for P01 and P02 appear in R01; those for P03 and P04 in R02.
2. The corn from all four cropping activities is entered in the same transfer row. Oats from both P03 and P04 are transferred in the same row (R05) and standing meadow from both P03 and P04 is transferred in R07.
3. The standing meadow in P03 and P04 is transferred in the same R07 transfer row. This assumes that the yield and quality of one acre of meadow in a CCOM sequence is comparable to one acre of meadow in a CCOMM rotation. Where this assumption is not

realistic, separate transfer rows can be provided for each type of meadow.

MODEL 3.8: *Use of Multiple Growing Activities, Multiple Transfer Rows, and Multiple Harvesting Activities to Accommodate Several Classes of Land*

Explanation

This model illustrates a structure which permits alternate harvesting activities when several classes of land are cropped, each having different yield potential. The structure of this model parallels that of Models 3.4 and 3.7.

P01 = a continuous corn raising activity on Class I land. The unit of activity is one acre.
P02 = a corn-corn-soybean growing activity on Class I land. The unit of activity is three acres.
P03 = a CCOM growing activity only on Class II land. The activity unit is four acres.
P04 = a corn harvesting activity on Class I land. The unit of activity is one acre.
P05 = a corn harvesting activity on Class II land. The unit of activity is one acre.
P06 = a corn silage harvesting activity on Class I land. The activity unit is one acre.
P07 = a corn silage harvesting activity on Class II land. The activity unit is one acre.
R01 = a Class I land restraint. Each unit in the B column is one acre.
R02 = a Class II land restraint. Each B column unit represents one acre.
R03 = a labor restraint. Each B column unit is one acre.

			Real Activities						
Row Type		B	P01	P02	P03	P04	P05	P06	P07
N	C		—27	—70	—67	—6.5	—6.25	—18	—17
L	R01	240	1	3					
L	R02	80			4				
L	R03	2,400	5	14	12	1.2	1.1	5.3	5.0
L	R04		—1	—2		1		1	
L	R05				—2		1		1
L	R06					—90	—80		
L	R07				—1				
L	R08				—1				
L	R09			—1					
L	R10							—16	—14

R04 = a row transferring corn produced on Class I land. The transfer unit is one acre.
R05 = a row transferring corn produced on Class II land. The transfer unit is one acre.
R06 = a corn grain transfer row. The transfer unit is one bushel.
R07 = a standing oats transfer row. The transfer unit is one acre.
R08 = a standing meadow transfer row. The transfer unit is one acre.
R09 = a standing soybean transfer row. The transfer unit is one acre.
R10 = a corn silage transfer row. The transfer unit is one ton.

Points to Observe

1. The model is incomplete because it does not provide for feeding or selling corn and silage, nor does it include harvesting activities for soybeans, oats, and meadow.
2. The transfer unit for rows R04 and R05 is in acres, but R06 transfers bushels.
3. Two rows transferring standing corn (R04 and R05) are needed because the yield derived from growing activities P01 and P02 differs substantially from P03. The standing corn to be produced on Class I land has the potential to yield 90 bushels of corn per acre compared with 80 bushels from Class II land.
4. Corn on Class I land yields 16 tons of silage and corn on Class II land only 14 tons.
5. Note that only one corn grain transfer row is needed. This indicates that a bushel of corn from Class I land is the same as a bushel from Class II land. Silage is transferred in a single row for the same reason.
6. A similar structure can be used where land is part owned and part rented on a crop share lease. In this case the standing corn from owned land would be transferred in rows separate from the corn on rented land.

MODEL 3.9: *Multiple Pasture Restraints*

Explanation

Sometimes part of the land, although unsuitable for row cropping, may be a potential source of pasture. Such land should be treated in a separate category in forming restraints. Often a single permanent pastureland restraint will suffice. However, where there are substantial differences in potential yield among areas, two or more pastureland rows are desirable. The model below illustrates a situation where permanent pastureland is divided into two classes, improvable and unimprovable.

P01 = a continuous corn growing and harvesting activity on Class I land. The unit of activity is one acre.
P02 = a CCOM activity on Class II land. The unit of activity is four acres.
P03 = a pasture improvement activity. The unit of activity is one acre.
P04 = an activity to provide for weed control and maintenance of fences on unimprovable pastureland. The unit of activity is one acre.
P05 = an activity which supplements permanent pasture with meadow from rotated land. The unit of activity is one acre.
P06 = a beef cow-calf production activity. The unit of activity is one cow.
P07 = a hay baling activity. The unit of activity is one acre.
R01 = a Class I land restraint. Each B column unit is one acre.
R02 = a Class II land restraint. Each B column unit is one acre.
R03 = a restraint on improvable permanent pasture. Each B column unit is one acre.
R04 = a restraint on unimprovable permanent pastureland. The unit of restraint is one acre.
R05 = a restraint on labor. The unit of restraint is one hour.
R06 = a standing corn transfer row produced on Class I land. The unit of transfer is one acre.
R07 = a standing corn transfer row produced on Class II land. The unit of transfer is one acre.
R08 = a standing oats transfer row. The transfer unit is one acre.
R09 = a standing meadow transfer row. The transfer unit is one acre.
R10 = a standing pasture transfer row. The unit of transfer is one ton of hay equivalent.
R11 = a hay transfer row. The unit of transfer is one ton.

			Real Activities						
Row Type		B	P01	P02	P03	P04	P05	P06	P07
N	C		—32	—82	—14.00	—2.00	—2.00	88	—14.00
L	R01	240	1						
L	R02	80		4					
L	R03	80			1				
L	R04	40				1			
L	R05	2,400	5	16	1.3	.8	.2	8.0	6
L	R06		—1						
L	R07			—2					
L	R08			—1					
L	R09			—1			1		1
L	R10				—2.6	—1.4	—3.4	3.8	
L	R11							1.8	—3.6

Points to Observe

1. The number of acres of each class of land available is determined and entered in the B column of the appropriate restraint row. In the model presented there are 240 acres of Class I land, 80 acres of Class II land, 80 acres of improvable pastureland, and 40 acres of unimprovable pastureland.
2. P03, the pasture improvement activity, has a labor coefficient at its intersection with the labor row, R05. The cost of the original improvement is amortized and included in the C row with the yearly expenditures necessary to maintain the improved pasture.
3. The C row coefficient for P04 should reflect the allocable costs incurred in clipping weeds and maintaining fences. Labor required for such maintenance is entered at the intersection of P04 and the labor row, R05.
4. P03 and P04 require an acre of pastureland and transfer pasture feed to the standing pasture transfer row. The coefficients in R10 represent tons of hay equivalent in the form of pasture. Note that the P03 coefficient in R10 is —2.6 while the P04 coefficient in R10 is —1.4. The difference is explained by the higher yield potential of the improved pasture. In forming these coefficients one can start with high-quality rotation pasture as a bench mark. One would expect a yield of 3–4 tons of hay equivalent from such pasture. Coefficients for pasture requirements of livestock activities must also be stated in terms of tons of pasture equivalent to be consistent with the pasture supplying coefficients in R10.

MODEL 3.10: *Alternative Production Methods*

Explanation

One may wish to compare different methods of producing a crop or livestock product. Examples might be (1) corn produced with different levels of fertilization; (2) minimum tillage systems versus conventional methods; or (3) variations in the schedule of plowing, planting, and harvesting crops. Each variation in method, level of fertilization, or scheduling of operations for each crop can be identified as a distinct activity.

P01 = a continuous corn growing and harvesting activity with a low level of fertilizer use. The unit of activity is one acre.

P02 = a continuous corn growing and harvesting activity with a medium level of fertilizer use. The unit of activity is one acre.

P03 = a continuous corn growing and harvesting activity with a high level of fertilizer use. The unit of activity is one acre.
R01 = a land restraint. The B column unit is acres.
R02 = a labor restraint. The B column unit is hours.
R03 = a corn transfer row. The transfer unit is bushels.

			Real Activities		
Row Type		B	P01	P02	P03
N	C		—24	—32	—36
L	R01	320	1	1	1
L	R02	2,400	4.8	5.0	5.1
L	R03	0	—80	—90	—94

Points to Observe

1. The differences among coefficients for P01, P02, and P03 in R03 reflect the expected response from fertilizer use; the net price variations observable in the C row reflect the differences in variable costs associated with the three levels of fertilizer use.
2. This model contains no capital restrictions. If the model included one or more capital restraints, the differences in the amount of capital required for the three levels of fertilizer use would be reflected in the P01, P02, and P03 coefficients in the capital row.

MODEL 3.11: *Buying and Selling Activities*

Explanation

The B column value specified in any resource restraint or transfer row may be supplemented by the addition of a purchasing activity.

P01 = a continuous corn growing and harvesting activity.
P02 = a corn selling activity. The unit of activity is one bushel.
P03 = a hog farrowing, finishing, and marketing activity. The unit of activity is one litter farrowed in February and a second in August.
P04 = a corn buying activity. The activity unit is one bushel.
R01 = a land restraint. The B column unit is acres.
R02 = a labor restraint. The B column unit is hours.
R03 = a corn transfer row. The transfer unit is one bushel.

			Real Activities			
Row Type		B	P01	P02	P03	P04
N	C		—32	1.20	395	—1.30
L	R01	320	1			
L	R02	2,400	5		32	
L	R03	0	—90	1	190	—1

Points to Observe

1. The P04 (corn purchasing) coefficient in R03 (corn transfer) carries a negative sign, indicating that the purchasing activity will add to the R03 supply. The appropriate sign becomes clearer if we recall that each row in the model represents an equation after disposal activities have been added. The B column represents the left side of the equation, the activities and their respective coefficients its right side. When the (P04) (—1) product is transposed, it becomes a positive number in the B column of R03.
2. The coefficient for P02 (corn selling) in R03 (corn transfer) is positive.
3. The C row entry for P04 (corn buying) is negative because this activity of itself would not add to the value of the program but constitutes a cost not accounted for elsewhere in the model. The C row entry for P02 (corn selling) is positive. The differences between the absolute net prices of P02 and P04 (i.e., ten cents per bushel) reflect transportation and other marketing costs associated with buying and selling corn.
4. The model provides that the corn raised on the farm may be either fed to hogs or sold, whichever is more profitable. Furthermore, in this model corn supplies available for hog production may be supplemented by purchased feed grains, providing for an expansion in hog numbers beyond the home-raised feed supply if such is consistent with the most profitable use of all resources.

MODEL 3.12: *Subjective Restraints*

Explanation

An operator may impose subjective restraints on the level at which an activity can enter the program. These may be maximum or minimum restraints. This model considers only maximum restraints. The use of minimum restraints will be illustrated later. Activities may be restrained as a protection against risk or simply because the operator has a strong preference for limiting the time and attention he gives to a particular activity.

P01 = a corn growing and harvesting activity. The unit of activity is one acre.
P02 = a cattle buying, feeding, and selling activity. The unit of activity is one yearling steer.
P03 = a hay buying activity. The unit of activity is one ton.
R01 = a land restraint. The B column entry represents acres.
R02 = a labor restraint. The B column entry represents hours.

R03 = a subjective restraint on the number of cattle fed. The B column entry is number of head.
R04 = a corn transfer row. The transfer unit is one bushel.
R05 = a hay transfer row. The transfer unit is one ton.

Row Type		B	P01	P02	P03
N	C		—32	125	—17
L	R01	320	1		
L	R02	2,400	5	4	
L	R03	100		1	
L	R04	0	—90	42	
L	R05	0		1.4	—1

Points to Observe

1. The P02 coefficient in R03 (the subjective restraint) is +1.
2. The maximum level at which P02 can enter the solution is 100, the quantity specified by the B column entry in R03.
3. The C row entry for P02 is determined by estimating the gross sale price per finished steer and deducting the original cost of the steer and the allocable costs not accounted for in the model such as commercial feed purchased and veterinary expenses. Note that feed grain and hay consumption are accounted for in the model and hence are not deducted from the gross sale price in determining the C row entry.

MODEL 3.13: *Subjective Restraints Using the Bounding Feature*

Explanation

All activities below are the same as in Model 3.12. This model presents a system of forming maximum restraints on activities which is optional with some computing routines. Restraint R03 is eliminated and the restraint level, 100, from the B column in the above model is entered below in a row called "upper bound."

Row Type		B	P01	P02	P03
N	C		—32	125	—17
L	R01	320	1		
L	R02	2,400	5	4	
L	R04	0	—90	42	
L	R05	0		1.4	—1
	Upper Bound			100	
	Lower Bound			0	

Points to Observe

1. The formation of the maximum restraint in this manner reduces by one the number of rows within the matrix. This may be important in computer systems with limited capacity.
2. Both systems of restraining P02, illustrated in Model 3.12 and here, lead to exactly the same solution. In this illustration the choice between the two rests solely on convenience.
3. With multiple-feeding activities, a row restraint allows more competition between the activities.

MODEL 3.14: *Transfer of Young Stock to Selling, Feeding, or Replacement Activities*

Explanation

The question often arises in farm planning whether the commercial stock raiser should retain young stock he produces for his own herd or feedlot or sell them as feeders. One feature of the model below is designed to analyze this problem. The model also contains two activities, P07 and P09, which address the decision of producing versus buying replacement stock.

P01 = a corn-oats-meadow rotation growing activity. The unit of activity is three acres.
P02 = a beef cow-calf production activity. The unit of activity is one cow producing .9 calf (.45 of a 400-pound heifer and .45 of a 450-pound steer).
P03 = a heifer calf selling activity. The unit of activity is one 400-pound calf.
P04 = a steer calf selling activity. The unit of activity is one 450-pound steer calf.
P05 = a heifer feeding and selling activity. The unit of activity is one 400-pound heifer fed to 1,000 pounds.
P06 = a steer feeding and selling activity. The unit of activity is one 450-pound steer fed to 1,100 pounds.
P07 = a heifer growing and breeding activity for the replacement of cull cows. The activity unit is one heifer.
P08 = a cull cow selling activity. The unit of activity is one cow.
P09 = a replacement heifer buying activity. The unit of activity is one replacement heifer.
R01 = a maximum restraint on labor. The B column unit is one hour.
R02 = a maximum restraint on pastureland. The unit of restraint is one acre.
R03 = a Class I land restraint. Each B column unit is one acre.
R04 = a standing meadow transfer row. The transfer unit is one acre.

MODEL 3.14

Row Type		B	P01	P02	P03	P04	P05	P06	P07	P08	P09
N	C		—78.00	—24.00	108.00	130.50	238	285	—28	180	—240
L	R01	2,400	9.0	11.00			6.0	7.5	4.0		
L	R02	120		2.30					.6		
L	R03	300	3								
L	R04		—1								
L	R05		—1								
L	R06		—1								
L	R07	75		1.00							
L	R08			—.45	1		1		1		
L	R09			—.45		1		1			
L	R10			—.16						1	
L	R11			.16					—1		—1
L	R12			1.80			1.1	1.0	1.3		
L	R13			5.00			42	60	20		

R05 = a standing corn transfer row. The transfer unit is one acre.
R06 = a standing oats transfer row. The transfer unit is one acre.
R07 = a maximum restraint on beef cow numbers. The unit of restraint is one cow.
R08 = a heifer calf transfer row. The transfer unit is one 400-pound heifer calf.
R09 = a steer calf transfer row. The unit of transfer is one 450-pound steer calf.
R10 = a cull cow transfer row. The unit of transfer is one 1,050-pound cow.
R11 = a replacement heifer transfer row. The transfer unit is one replacement heifer.
R12 = a hay transfer row. The unit of transfer is one ton.
R13 = a corn grain transfer row. The unit of transfer is one bushel.

Points to Observe

1. Defining the cow-calf activity unit as one cow results in the transfer of fractions of calves (assuming less than a 100% calf crop—half male and half female). The activity could be redefined with the unit of 100 cows, but the solution's activity level that results would be in hundreds of cows and must then be multiplied by 100 to obtain numbers of cows.
2. Activity P02 assumes a 90% calf crop. One half of the calves born are heifers and the other half bulls. Thus the coefficient at the intersection of P02 (the cow-calf activity) and R08 (the heifer calf transfer row) is —.45. The coefficient is the same for P02, R09. The latter is a steer calf transfer row because castration of bull calves is performed before weaning.
3. Sixteen percent of the cows are also culled each year, resulting in the —.16 in the cull cow transfer row. Assuming that the herd size is maintained at a constant level, a positive .16 appears in the replacement heifer transfer row to replace the .16 cull cow transferred in R10.
4. P03 and P04 permit selling the calves as feeders, and P05 and P06 represent feeding and selling for slaughter. The heifer calf selling activity assumes the sale of a 400-pound calf, and the heifer feeding and selling assumes a starting weight of 400 pounds. Likewise, steer calf selling assumes the sale of a 450-pound steer calf, and the feeding and selling activity has a starting weight of 450 pounds.
5. The C row coefficients for the feeding and selling activities may seem high, but the reader should note that no purchase price has been subtracted from gross returns because the feeder animals originated from the cow-calf activity within the model.

6. P08 is included in the model to provide for selling cull cows. Another approach is to delete R10 from the model and add the return from selling .16 cows to the C row coefficient of P02.
7. P07 provides the alternative of raising replacements on the farm for the cows culled from the herd. The coefficient at the intersection of P07 (heifer raising activity) and R08 (heifer calf transfer row) is +1. The P07 coefficient in R11 (replacement heifer transfer) is —1 since P07 adds to this row. P07 requires a heifer calf plus feed, labor, and other expenses unaccounted for in the model to grow and breed a replacement heifer.
8. P09 provides for purchasing replacement stock. Because this activity makes no demands on feed, pasture, and labor, zero coefficients appear in these rows.
9. A steer calf purchasing activity could be added to the model by entering the appropriate C row coefficient (purchase price plus transportation costs) and a —1 in the steer calf transfer row. Likewise, a heifer calf purchasing activity could be added to supplement the heifer feeding and selling activity.
10. The pattern illustrated in this model could be extended with few modifications to the culling and replacement process with swine or dairy activities.

MODEL 3.15: *Transfer of Livestock by Weight*

Explanation

The question of how the optimum plan would be influenced by changes in the ratios of livestock prices frequently arises in farm planning. The model illustrated below is similar to Model 3.14 but is structured to facilitate analysis of multiple sets of price relationships through use of the multiple C row option or to simplify changing price coefficients from one run to another.

P01 = a corn-oats-meadow growing activity. The unit of activity is three acres.
P02 = a beef cow-calf production activity. The unit of activity is one cow producing .9 calf (.45 of a 400-pound heifer and .45 of a 450-pound steer).
P03 = a 400-pound heifer calf selling activity. The unit of activity is one hundredweight.
P04 = a 450-pound steer calf selling activity. The unit of activity one hundredweight.
P05 = a heifer feeding activity. The unit of activity is one 400-pound heifer fed to 1,000 pounds.

P06 = a steer feeding activity. The unit of activity is one 450-pound steer fed to 1,100 pounds.
P07 = a heifer growing and breeding activity for the replacement of cull cows. The activity unit is one heifer.
P08 = a cull cow selling activity. The unit of activity is one cow.
P09 = a replacement heifer buying activity. The unit of activity is one replacement heifer.
P10 = a 1,000-pound slaughter heifer selling activity. The unit of activity is one hundredweight.
P11 = a 1,100-pound slaughter steer selling activity. The unit of activity is one hundredweight.
R01 = a maximum restraint on labor. The B column unit is one hour.
R02 = a maximum restraint on pastureland. The unit of restraint is one acre.
R03 = a Class I land restraint. Each B column unit is one acre.
R04 = a standing meadow transfer row. The transfer unit is one acre.
R05 = a standing corn transfer row. The transfer unit is one acre.
R06 = a standing oats transfer row. The transfer unit is one acre.
R07 = a maximum restraint on beef cow numbers. The unit of restraint is one cow.
R08 = a 400-pound heifer calf transfer row. The transfer unit is pounds.
R09 = a 450-pound steer calf transfer row. The transfer unit is pounds.
R10 = a cull cow transfer row. The transfer unit is one 1,050-pound cow.
R11 = a replacement heifer transfer row. The transfer unit is one replacement heifer.
R12 = a hay transfer row. The transfer unit is one ton.
R13 = a corn grain transfer row. The transfer unit is one bushel.
R14 = a 1,000-pound slaughter heifer transfer row. Each unit of transfer is pounds.
R15 = a 1,100-pound slaughter steer transfer row. Each unit of transfer is pounds.

Points to Observe

1. The C row elements for the selling activities (P03, P04, P10, and P11) are set equal to the hundredweight sale price less marketing costs per hundred pounds.
2. Separate transfer rows were constructed for the steer and heifer calves because of differences in their selling price and their response in the feedlot. If alternatives were given for feeding 500-pound heifers and 550-pound steers, separate transfer rows for each must be constructed for transfer of the calves from the buying to feeding activities. However, if the slaughter weights, carcass grades, and expected selling price are comparable, no additional rows for transfer of slaughter animals are needed.

MODEL 3.15

Row Type		B	P01	P02	P03	P04	P05	P06	P07	P08	P09	P10	P11
N	C		—78.00	—24	29.00	31.00	—19.15	—20.50	—28	180	—240	27.00	30.00
L	R01	2,400	9.0	11			6.0	7.5	4.0				
L	R02	120		2.3					.6				
L	R03	300	3										
L	R04		—1										
L	R05		—1										
L	R06		—1										
L	R07	75		1									
L	R08			—180	100		400		400				
L	R09			—202.5		100		450					
L	R10			—.16						1			
L	R11			.16					—1		—1		
L	R12			1.8			1.1	1.0	1.3				
L	R13			5.0			42	60	20				
L	R14						—1,000					100	
L	R15							—1,100					100

MODEL 3.16: *Multiple Labor Restraints*

Explanation

The illustrations presented thus far have included a single labor restraint. Such a structure implies that labor can be freely substituted among seasons of the year. Obviously, labor during the winter months is likely to have an opportunity cost much different from the spring and early summer months. To plan realistically one must take account of the seasonality of labor requirements and restraints. Although labor restraints can be formed for every month of the year, this implies a rigidity in the timing of farming operation and consequently of labor use which also may be unrealistic. Restraints should be formed to focus on those periods of the year in which labor allocation is critical. The remaining noncritical periods also can be included to provide a complete accounting for labor within the system. In the Corn Belt April, May, and June typically constitute one crucial period and October and November another. Thus one can form four labor restrictions as follows: (1) April-May-June, (2) July-August-September, (3) October-November, and (4) December-January-February-March. The reader should not imply from this example that the periods must begin and terminate on the first or last of any month. One could form periods which extend from April 15 to June 20 or September 20 to November 15, if he felt periods defined in this manner corresponded more realistically to the pattern of critical labor periods.

P01 = a continuous corn growing and harvesting activity. The unit of activity is one acre.
P02 = a corn selling activity. The unit of activity is one bushel.
R01 = a land restraint. The B column entry is acres.
R02 = an April-May-June labor restraint. The B column entry is hours.
R03 = a July-August-September labor restraint. The B column entry is hours.
R04 = an October-November labor restraint. The B column entry is hours.
R05 = a December-January-February-March labor restraint. The B column entry is hours.
R06 = a corn transfer row. The transfer unit is one bushel.

Points to Observe

1. The B column entries imply that the operator would be willing to work long hours during critical seasons of the year but not during the entire year.

Row Type		B	P01	P02
N	C		—32	1.25
L	R01	320	1	
L	R02	730	2.8	
L	R03	520	.2	
L	R04	510	1.9	
L	R05	640	.1	
L	R06		—90	1

2. P01 has four labor coefficients, one for each season of the year. The P01 coefficient for R02 (April-May-June labor) is an estimate of labor required by one unit of P01 during these three months of the year. The coefficients in the rows designating the other time periods are interpreted similarly.

MODEL 3.17: *Timing of Livestock Production Activities*

Explanation

Frequently one wishes to investigate the optimum timing of production and marketing, especially with livestock activities. Two or more hog farrowing systems which provide for variations in the timing of farrowing should be treated as distinct activities. Hence a two-litter system on a January-July farrowing schedule becomes one activity, and a two-litter system on a February-August schedule becomes another.

P01 = a continuous corn growing and harvesting activity. The unit of activity is one acre.
P02 = a hog farrowing, finishing, and selling activity on a January-July schedule. The unit of activity is two litters, one farrowed in January and one in July.
P03 = a hog farrowing, finishing, and selling activity on a February-August schedule. The unit of activity is two litters, one farrowed in February and one in August.
R01 = a land restraint. The B column entry is acres.
R02 = an April-May-June labor restraint. The B column entry is hours.
R03 = a July-August-September labor restraint. The B column entry is hours.
R04 = an October-November labor restraint. The B column entry is hours.
R05 = a December-January-February-March labor restraint. The B column entry is hours.
R06 = a corn transfer row. The transfer unit is bushels.

Row Type		B	Real Activities P01	P02	P03
N	C		—32	410	395
L	R01	320	1		
L	R02	730	2.8	8	8
L	R03	520	.2	9.5	8
L	R04	510	1.9	6.5	7
L	R05	640	.1	11	9.5
L	R06		—90	195	190

Points to Observe

1. The P02 and P03 activities have different net prices and coefficients, hence they are separate activities. The differences in labor requirements are important in the development of the optimum farm plan.
2. The hog producing and selling activities (P02 and P03) have coefficients in the corn transfer row. Hence the cost of corn should not be deducted (included as a variable cost) in estimating the C row coefficients for P02 and P03.

MODEL 3.18: ***Restraints on Availability of Hired Labor***

Explanation

Multiple labor restraints may be supplemented by labor hiring activities. In turn, there may be restraints on the amount of labor the operator can hire, and the amount may vary with seasons. Such a situation often exists where the potential supply of hired labor consists of part-time workers. Limiting the amount of labor which can be hired requires that an additional set of restraints be built into the model.

P01 = a corn growing and harvesting activity. The unit of activity is one acre.
P02 = a corn selling activity. The unit of activity is one bushel.
P03 = an April-May-June labor hiring activity. The unit of activity is one hour.
P04 = an October-November labor hiring activity. The unit of activity is one hour.
R01 = a land restraint. The B column entry is acres.
R02 = an April-May-June labor restraint. The B column entry is hours.
R03 = a July-August-September labor restraint. The B column entry is hours.
R04 = an October-November labor restraint. The B column entry is hours.

R05 = a December-January-February-March labor restraint. The B column entry is hours.
R06 = an April-May-June restraint on hired labor. The B column entry is hours.
R07 = an October-November restraint on hired labor. The B column entry is hours.
R08 = a corn transfer row. The transfer unit is bushels.

			Real Activities			
Row Type		B	P01	P02	P03	P04
N	C		–32	1.25	–1.75	–1.75
L	R01	320	1			
L	R02	730	2.8		–1	
L	R03	520	.2			
L	R04	510	1.9			–1
L	R05	640	.1			
L	R06	100			1	
L	R07	100				1
L	R08	0	–90	1		

Points to Observe

1. A separate labor hiring activity is required for each labor period supplemented by hired labor. P03 supplements the R02 (April-May-June) supply, and P04 supplements the R04 (October-November) supply. Because the hiring activities add to the supply in the R02 and R04 rows, their coefficients are negative.
2. R06 restricts the amount of labor that can be hired during April-May-June to a maximum of 100 hours. Similarly, R07 restricts labor hiring during October and November to 100 hours. The P03 coefficient in R06 is positive, because each hour hired reduces the B column value in R06 by one hour. The same interpretation holds for the P04 coefficient in R07.
3. Net prices for P03 and P04 are negative because the labor hiring activities themselves subtract from the value of the program. The negative sign on these two activities is the opposite of P02 (the corn selling activity) because the latter adds to the value of the program.
4. The bounding option where available can be used to restrain the level of P03 and P04. In this case rows R06 and R07 are eliminated from the model and upper bounds of 100 imposed on both P03 and P04.

MODEL 3.19: *Minimum Restraints*

Explanation

All restraints encountered thus far have been maximum restraints. They restrict activities to no more than some specified level. Minimum restraints require that activities enter the plan equal to or greater than some minimum level. Thus, while a maximum restraint on a dairy activity limits the number of cows in the plan to *no more than* the maximum level of activity of (say) 40, a minimum restraint assures that the level of the dairy activity will enter the plan at no less than (say) 20 cows. In this model all disposal activities are included to emphasize the difference between minimum and maximum restrictions. The artificial activity essential for the minimum restraint is also included. In most computing routines appropriate disposal and artificial activities are supplied automatically, once the restriction has been properly identified. Note that the row type label given the minimum restraint is "G" to signify an *equal to or greater than* restraint.

P01 = an activity which includes growing and harvesting corn and oats and growing meadow in a CCOM sequence. The unit of activity is four acres including two of corn, one of oats, and one of meadow.
P02 = a dairy activity. The unit of activity is one cow.
P03 = a haymaking activity. The unit of activity is one acre.
R01 = a land restraint. The B column entry is acres.
R02 = a labor restraint. The B column entry is hours.
R03 = a minimum restraint on the level of the dairy activity. The B column entry is number of cows.
R04 = a corn transfer row. The transfer unit is bushels.
R05 = an oats transfer row. The transfer unit is bushels.
R06 = a standing meadow or pasture transfer. The transfer unit is one acre.
R07 = a hay transfer row. The transfer unit is tons.

			Real Activities			Disposal Activities							Artificia Activity
Row Type		B	P01	P02	P03	R01	R02	R03	R04	R05	R06	R07	R03A
N	C		—86	280	—16	0	0	0	0	0	0	0	—M
L	R01	320	4			1							
L	R02	2,400	16	80	2		1						
G	R03	20		1				—1					1
L	R04	0	—185	30					1				
L	R05	0	—62	0	0					1			
L	R06	0	—1	2.5	1						1		
L	R07	0		3	—3.1							1	

Points to Observe

1. Because column R03 represents a disposal activity for a minimum restraint, its coefficient in row R03 is —1. Coefficients for all disposal activities on maximum restraints are +1.
2. All disposal activities have a zero net price.
3. An artificial activity is included for the minimum restraint. Its coefficient is +1 in the R03 (minimum restraint) row.
4. The net price on the artificial activity is some very high negative value represented by —M. The artificial activity provides an artificial solution from which the simplex procedure may proceed toward an optimum solution. The high negative price on the artificial activity assures that it will leave the basis at an early iteration in the simplex procedure.
5. Specification of a minimum restraint on any activity such that one or more maximum restraints must be exceeded cannot of course result in a solution. In this model, for example, specifying a minimum of 35 dairy cows instead of 20 would not be consistent with the labor restraint. Thirty-five cows would require 2,800 hours of labor since the P02 coefficient in R02 is 80, exceeding the B column value in R02 of 2,400 hours. Structuring such infeasibilities into a model can often be avoided by checking the requirements of the activity that is to be forced to some minimum level and ascertaining whether any B column entries will be exceeded at the minimum level imposed. Inconsistencies are not readily apparent, particularly where B column values are supplemented or are generated entirely by other activities in the model. However, there is sufficient possibility of identifying inconsistencies to justify a careful check prior to computation.
6. The bounding option also can be used to impose minimum restraints. Thus R03 would be deleted from the model and a lower bound of 20 entered in an additional row formed at the bottom of the model and labeled "lower bound."

MODEL 3.20: *Minimum and Maximum Restraints*

Explanation

An activity may be restrained to both a minimum and a maximum level. Such a procedure might be appropriate where the planner has an attachment to dairying and wishes to assure himself that a herd of at least 20 cows will be included in the final plan. Assuring that this condition will be met requires a minimum restraint. On the other hand, facilities might limit the size of herd to a maximum of 40 cows, necessitating a maximum restraint. With both minimum and maximum restraints specified in the model, the dairy herd would enter the plan at a level of at least 20 but no more than 40 cows.

P01 = an activity including growing and harvesting corn and oats and growing meadow in a CCOM sequence. The unit of activity is four acres.

			Real Activities			Disposal Activities								Artificial Activity
Row Type		B	P01	P02	P03	R01	R02	R03	R04	R05	R06	R07	R08	R03A
N	C		−86	280	−16	0	0	0	0	0	0	0	0	−M
L	R01	320	4			1								
L	R02	2,400	16	80	2		1							
G	R03	20	0	1				−1						1
L	R04	40	0	1					1					
L	R05	0	−185	26						1				
L	R06	0	−62	8							1			
L	R07	0	−1	2.5	1							1		
L	R08	0		3	−3.1								1	

P02 = a dairy activity. The unit of activity is one cow.
P03 = a haymaking activity. The unit of activity is one acre.
R01 = a land restraint. The B column entry is acres.
R02 = a labor restraint. The B column entry is hours.
R03 = a minimum dairy restraint. The B column entry is number of cows.
R04 = a maximum capacity restriction on the level of the dairy activity. The B column entry is number of cows.
R05 = a corn transfer row. The transfer unit is bushels.
R06 = an oats transfer row. The transfer unit is bushels.
R07 = a standing meadow transfer row. The transfer unit is acres.
R08 = a hay transfer row. The transfer unit is tons.

Points to Observe

1. The coefficient for the disposal activity for the minimum restraint (row R03) is —1 in the R03 column.
2. All disposal activities have a zero net price.
3. The minimum restraints require an artificial activity labeled R03A above. This activity has a +1 in row R03 (minimum restraint).
4. The dairy activity P02 has a coefficient of +1 both in row R03 and in row R04.
5. The bounding option can be used to impose both a minimum and a maximum restraint on P02. In this case R03 and R04 are deleted and an upper bound of 40 and a lower bound of 20 are formed.

MODEL 3.21: *Equality Restraints*

Explanation

Maximum and minimum restraints have been illustrated thus far. In this model we deal with a third type of restraint—the equality restraint. Use of equality restraints permits one to specify that an activity must enter the solution at a predetermined level. In the case of dairy cows the planner may wish to provide for exactly 25 cows, no more and no less, regardless of their profitability relative to alternative activities. The equality restraint has no disposal activity since neither under- nor overfulfillment of the B column entry in the equality restraint row is permitted. An artificial activity is required to provide a starting basis for the simplex routine.

P01 = a dairy activity. The unit of activity is one cow.
R02A = an artificial activity.
R01 = a labor restraint. The B column entry is hours.
R02 = an equality restraint. The B column entry is number of cows.
R03 = a corn transfer row. The transfer unit is bushels.

R04 = a standing meadow transfer row. The transfer unit is acres.

R05 = a hay transfer row. The transfer unit is tons.

			Real Activity	Disposal Activities				Artificial Activity
Row Type		B	P01	R01	R03	R04	R05	R02A
N	C		280	0	0	0	0	—M
L	R01	2,400	80	1		.		
E	R02	25	1					1
L	R03	0	30		1			
L	R04	0	2.5			1		
L	R05	0	3				1	

Points to Observe

1. In contrast to the other restraints, no disposal activity is provided for the equality R02.
2. R02A is an artificial activity with a coefficient of +1 in row R02.
3. The C row entry for R02A is —M, a very high negative quantity.
4. As in the case of minimum restraints, with most computing routines it is not necessary to enter artificial activities for equality restraints. Note that the equality restraint is identified by the letter E in the row type column.
5. The equality can also be formed through use of the bounding feature. In this case a fixed bound of 25 would be established on P01, the dairy activity.

MODEL 3.22: *Fixed Cost Accounting Activities*

Explanation

Equality restraints can also be utilized to provide one or more "fixed cost paying" activities in the model. Where all nonallocable or fixed costs involved in the operation are accounted for in this manner, the value of the program resulting from the maximization process will be an estimate of net income. If one wishes to provide an estimate of capital accumulation, he can also include an activity to deduct estimates of family living expenses.

The reader should note that fixed costs can be subtracted readily from the value of the program after the model has been optimized. Ordinarily this is a lower cost method of arriving at an estimate of net income than including fixed cost activities and equality restraints in the model. However, in some types of dynamic models where the capital supply in the second and subsequent planning periods is a function of capital accumulation in previous periods, the use of activities to deduct fixed costs and family living expenses becomes necessary.

P01 = a corn growing and harvesting activity. The unit of activity is one acre.
P02 = a fixed cost paying activity. The activity unit is the estimated amount of fixed costs.
R04A = an artificial activity.
R01 = a land restraint. The B column entry is acres.
R02 = a labor restraint. The B column entry is hours.
R03 = a corn transfer row. The transfer unit is bushels.
R04 = an equality restraint. The B column unit represents one unit of fixed cost. In this case the unit of fixed cost equals $9,000.

			Real Activity	Disposal Activities				Artificial Activity
Row Type		B	P01	P02	R01	R02	R03	R04A
N	C		–32	–9,000	0	0	0	–M
L	R01	320	1		1			
L	R02	2,400	5			1		
L	R03	0	–90				1	
E	R04	1		1				1

Points to Observe

1. R04 (equality restraint) has no disposal activity but has a coefficient in column R04A (artificial activity).
2. The net price entry for P02 (fixed cost paying activity) is negative and is the estimated total fixed costs.
3. An alternate structuring of the model will accomplish the same purpose and may seem more straightforward. Define P02 as a fixed cost paying activity where the activity unit is $1 and the net price is –$1. Define the B column units as equal to $1 per unit. Thus the entry in the B column becomes 9,000–the amount of fixed cost to be paid. P02 thus is restricted to entering only at a level of $9,000.

MODEL 3.23: *Multiple B Columns*

Explanation

All models previously illustrated have been designed to provide a single optimization. Frequently it is useful to explore the outcome if one or more resources were restrained at more than one level. To illustrate, one may be interested in the optimum plan and expected income if 320 instead of 240 acres of Class I land were available. This information could be obtained from successive models involving two or more computer runs. The multiple B column model provides the same results with little more time and effort than is required for a single solution.

P01 = a continuous corn growing and harvesting activity on Class I land. The unit of activity is one acre.
P02 = a corn-corn-soybean growing and harvesting activity on Class I land. The unit of activity is three acres.
P03 = a CCOM activity on Class II land. The unit of activity is four acres.
P04 = a CCOMM activity on Class II land. The unit of activity is five acres.
B = a column specifying the restraint limits.
B2 = a column specifying a second set of restraint limits.
R01 = a Class I land restraint. The B column unit is acres.
R02 = a Class II land restraint. The B column unit is acres.
R03 = a labor transfer row. The B column unit is hours.
R04 = a grain transfer row. The transfer unit is bushels.
R05 = an oats transfer row. The transfer unit is bushels.
R06 = a soybean transfer row. The transfer unit is bushels.
R07 = a standing meadow transfer row. The transfer unit is acres.

				Real Activities			
Row Type		B	B2	P01	P02	P03	P04
N	C			–32	–84	–82	–86
L	R01	240	320	1	3		
L	R02	80	80			4	5
L	R03	2,400	2,400	5	14	16	19
L	R04	0	0	–90	–178	–185	–190
L	R05	0	0			–60	–60
L	R06	0	0		–30		
L	R07	0	0			–1	–2

Points to Observe

1. This model except for the multiple B column features is identical to Model 3.7.
2. Only the B column entry for Class I land is changed from B to B2. All other entries are repeated in B2.
3. Any or all of the entries in column B2 may differ from those in B.
4. This procedure results in two optimizations, although only one computer run is required.

5. Any number of B columns may be built into the model. Thus the marginal value product may be estimated for a resource restraint at any level desired by defining additional B columns with appropriate B column entries.

MODEL 3.24: *Multiple Objective Functions*

Explanation

This model is identical to Model 3.3 except that the concept of a multiple objective function is introduced. The latter permits two solutions based on differing price expectation from a single computer run.

P01 = a corn growing and harvesting activity. The unit of activity is one acre.
P02 = a corn selling activity. The unit of activity is one bushel.
P03 = a hog raising and selling activity under a one-farrowing-per-year system. The unit of activity is one sow and a litter of seven pigs.
R01 = a land restraint. The B column unit is acres.
R02 = a labor restraint. The B column unit is hours.
R03 = a corn transfer row. The transfer unit is one bushel.
C = one objective function.
C2 = a second objective function where price relationships have been altered.

Row Type		B	P01	P02	P03
N	C		–32	1.20	205.20
N	C2		–32	1.40	205.20
L	R01	320	1		
L	R02	2,400	5		16
L	R03	0	–90	1	104

Points to Observe

1. A model with two objective functions results in two optimizations.
2. Any or all coefficients in the C2 row may differ from its counterpart in the original C row.
3. Changing one or more coefficients but not all has the effect of altering the price relationships and hence the competitive position of the activities.
4. Any number of objective functions may be structured into the model.

MODEL 3.25: *Models Containing Both Multiple B Columns and Multiple C Rows*

Explanation

Multiple B columns and multiple C rows can be combined into a single model. Again the same information can be obtained from successive solutions of simpler models, each containing only one B column and one C row. The purpose of the multiple B column and multiple C row combination is to save effort and computer time where several optimizations are contemplated.

P01 = a corn growing and harvesting activity. The unit of activity is one acre.
P02 = a corn selling activity. The unit of activity is one bushel.
R01 = a land restraint. The B column unit is acres.
R02 = a labor restraint. The B column unit is hours.
R03 = a corn transfer row. The transfer unit is bushels.

Row Type		B	B2	P01	P02
N	C			—32	1.20
N	C2			—32	1.50
L	R01	320	400	1	
L	R02	2,400	2,600	5	
L	R03	0	0	—90	1

Points to Observe

1. The illustration above contains two B columns and two C rows. Four optimum solutions will result from the following combinations:

   Solution I — B and C
   Solution II — B2 and C
   Solution III — B and C2
   Solution IV — B2 and C2

2. Any number of B columns and C rows may be built into a model. The number of optimizations can become very large as the numbers increase. For example, four B columns and four C rows would result in a total of 16 solutions if all combinations are optimized.
3. Although the pattern is similar, models containing combinations of B columns and C rows require a set of control cards unique to each combination.

MODEL 3.26: *Programming Field Operations with Tractor Time Restraints*

Explanation

Two new concepts are introduced in this model:

1. Activities are disaggregated to a level where each field operation becomes a separate activity. Such a narrow definition of activities is sometimes useful in models where the focus of the programming analysis is on the power, machinery, and labor arrangement.
2. Provisions are made for restraining the amount of field work that can be accomplished during critical periods of the year. These restraints are built around the tractor units and the maximum amount of field time one should expect from each unit.

P01 = a fall plowing activity with a 3–4 bottom plow and a 50 hp tractor. The unit of activity is one acre.
P02 = a spring plowing activity with a 3–4 bottom plow and a 50 hp tractor. The unit of activity is one acre.
P03 = a fall plowing activity with a 5–6 bottom plow and a 100 hp tractor. The unit of activity is one acre.
P04 = a spring plowing activity with a 5–6 bottom plow and a 100 hp tractor. The unit of activity is one acre.
P05 = a disking activity with a 50 hp tractor and a 12′ tandem disk. The unit of activity is one acre.
P06 = a disking activity with a 100 hp tractor and a 16′ tandem disk. The unit of activity is one acre.
P07 = a corn planting activity with a four-row planter; fertilizer and herbicides are applied during planting. The unit of activity is one acre.
P08 = a corn planting activity with a six-row planter (30″ rows); fertilizer and herbicides are applied during planting. The unit of activity is one acre.
P09 = a corn growing activity. The unit of activity is one acre.
R01 = Sept. 16–Dec. 15 labor restraint. The B column entry is hours.
R02 = Dec. 16–Mar. 15 labor restraint. The B column entry is hours.
R03 = Mar. 16–June 15 labor restraint. The B column entry is hours.
R04 = June 16–Sept. 15 labor restraint. The B column entry is hours.
R05 = Sept. 16–Dec. 15 restraint on operation of 50 hp tractor. The B column entry is hours.
R06 = Dec. 16–Mar. 15 restraint on operation of 50 hp tractor. The B column entry is hours.
R07 = Mar. 16–June 15 restraint on operation of 50 hp tractor. The B column entry is hours.
R08 = June 16–Sept. 15 restraint on operation of 50 hp tractor. The B column entry is hours.

R09 = Sept. 16–Dec. 15 restraint on operation of 100 hp tractor. The B column entry is hours.
R10 = Dec. 16–Mar. 15 restraint on operation of 100 hp tractor. The B column entry is hours.
R11 = Mar. 16–June 15 restraint on operation of 100 hp tractor. The B column entry is hours.
R12 = June 16–Sept. 15 restraint on operation of 100 hp tractor. The B column entry is hours.
R13 = a land restraint. The B column entry is acres.
R14 = a plowing service transfer row. The transfer unit is acres of plowing.
R15 = a disking service transfer row. The transfer unit is acres of disking.
R16 = a planting service transfer row. The transfer unit is acres of planting.

Points to Observe

1. The model contains systems of both labor restraints and tractor time restraints. The latter may be built on the hours the tractor can perform effectively in the field during any period of time. A separate restraint (or, in the case of multiple time periods, a separate set of restraints) has been defined for each size of tractor.
2. The B column entries for tractor time take into account (a) the proportion of total time within the period during which field work is feasible, (b) time needed for maintenance and repair, and (c) the maximum amount of night-time work considered feasible.
3. In this example the activities have been narrowed so that corn growing receives plowing, disking, and planting service from separate activities. The activities could be extended to include cultivating and harvesting.
4. The model contains a transfer row for each type of field service. The service is transferred to the activity or activities which are potential users of the service. Although P09 (corn growing) is the only potential user of services provided in P01 to P08 in this example, many activities may compete for the same services.
5. A particular service such as disking may be supplied by more than one activity. Thus disking with a 50 hp unit and 12′ disk may be defined as one activity and disking with a larger disk and a 100 hp tractor as another.

PRECAUTIONS IN MODEL BUILDING

Experience with hundreds of students has shown that errors made by beginning programmers follow a pattern. Thus it is

MODEL 3.26

Row Type		B	P01	P02	P03	P04	P05	P06	P07	P08	P09
N	C		—.87	—.87	—.54	—.54	—.26	—.16	—.28	—.24	—19.00
L	R01	800	.73		.49						
L	R02	600									
L	R03	750		.73		.49	.22	.14	.29	.26	
L	R04	750									
L	R05	460	.72								
L	R06	50									
L	R07	560		.72			.22		.25	.22	
L	R08	544									
L	R09	460			.47						
L	R10	50									
L	R11	560				.47		.14			
L	R12	544									
L	R13	400									1
L	R14		—1	—1	—1	—1					1
L	R15						—1	—1			1
L	R16								—1	—1	1

possible to anticipate potentially troublesome areas where the student should exercise particular caution. They are as follows:

1. Carelessness in estimating and entering coefficients in the model. Frequently coefficients are not placed in the row or column intended. Giving coefficients the wrong sign is also a common mistake, particularly in the C row and in transfer rows.
2. Failure to maintain consistency in the definition of coefficients. Coefficients in all cells of a row must be formed on the basis of a common unit. Regardless of whether the unit is bushels, acres, hours, dollars, or tons all coefficients in the same row on both the B column side and the right-hand side must adhere to the row definition. Furthermore, once the unit of activity has been defined, all coefficients in the column must relate to that activity unit. If the activity unit is one acre, all requirements, yields, and the net price coefficients must also adhere to the acre definition given the activity unit.
3. Structuring of models which are unbounded. This difficulty sometimes arises out of faulty logic. More frequently, however, the unbounded solution is the result of omitting coefficients or inconsistent definition of coefficients. An unbounded solution occurs when an activity is sufficiently competitive to enter the solution and its level is unrestrained. A simple example of structuring an unbounded solution is a provision for purchasing a commodity for less than its selling price while requiring no capital, labor, or other inputs. Thus, if a corn buying activity is provided with a C row coefficient of (say) $1.05 and a corn selling activity is also included in the model with a C row coefficient of $1.06, the activity levels and the value of the program could proceed to infinity. Computing routines are designed to diagnose such errors and to terminate search for an optimum solution before large amounts of computer time are wasted.

   Some unbounded models are not easily recognized because several activities may be interacting to create the unbounded condition. However, the output report that results from attempting to optimize an unbounded model typically contains error messages which help in diagnosing the irregularities that created the problem. But even with the leads provided by the output report, an occasional obscure difficulty can be located only after a comprehensive review of the complete structure of the model, the signs, and the magnitude of the coefficients.

4. Structuring of infeasible solutions. Infeasible solutions result from defining restraints that are impossible to fulfill. They occur in models which contain equality or minimum restraints in association with maximum restraints. To illustrate, if a minimum restraint of (say) 75 cows is imposed on a beef cow-calf raising activity but the amount of meadow is restrained so that no more than enough feed to support 74 cows can be raised or purchased, the restraints cannot be met and the program cannot be optimized. Although the existence of an infeasibility will be established quickly by the computing routine, the characteristics of the model which produce the difficulty are sometimes difficult to diagnose. Both the frequency and complexity of infeasibilities grow markedly as the number of minimum and equality restraints included in the model increase.
5. Providing insufficient flexibility in the range of activities defined in the model. This difficulty arises most frequently with cropping activities which define rotation sequences of crops. Suppose CCOM and CSBCOM (corn-soybean-corn-oats-meadow) growing activities are the only two cropping activities defined. In addition, a maximum and a minimum are defined for the percentage of cropland in each crop. With 100 acres of cropland no more than 50 or no less than 40 acres of corn can be grown and still utilize all of the cropland, or no more than 25 or no less than 20 acres of meadow can be produced. In models structured so inflexibly, the range of alternative cropping patterns considered may inadvertently be smaller than the planner intended. More often difficulty arises when such narrow specifications of alternatives are combined with restraints to meet subjective standards imposed by the operator or landlord or to analyze government program alternatives.
6. Often the programmer will restrict the amount of labor to be hired to the amount used in the existing program. This is too restrictive where additional labor can be hired, since it may not permit adequate testing of labor-intensive activities not in the current program or increasing the scale of the enterprise or enterprises presently in the farm plan.

CHAPTER FOUR

PROCEDURES FOR DATA PREPARATION AND PROCESSING

Computers are essential to the application of linear programming to realistic farm planning problems. Equally important is the computing system which guides the manipulation of the problem toward a solution. Preparation of such routines is a task requiring knowledge of linear programming methods and skill in programming computers. Fortunately, efficient systems have already been developed and tested. The task of the user is to familiarize himself with the capabilities and limitations of the routine, to learn how to prepare data for processing by it, and to develop the ability to interpret accurately and intelligently the output which results from the computing process.

PROGRAMMING SYSTEMS

The computing routine is often referred to loosely as a "program." It consists of a set of elementary instructions to which the computer can respond. These instructions may be stored on cards, tape, or disks. The process of programming routines to instruct the computer should not be confused with the analytical method, linear programming. Many different programs have been developed to instruct computers in a wide variety of data processing and analysis. You will be working with only one highly specialized computing routine (or system) among many that are utilized in a typical computing facility.

The same linear programming routine may be used to analyze a variety of problems. In one situation the user may seek to minimize the objective function, but on another occasion his problem may require a maximization procedure. Or he may want several optimizations involving several B columns and/or C rows. He also may have to choose among one or more options as to the type and form of output desired from the computing routine. Instructions of this nature are provided through the control cards which implement the system and cause the desired operations to be performed.

The data deck records all the coefficients from the model you have prepared. It provides labels for all rows and columns and specifies the magnitude of each coefficient included in the model.

Although most linear programming routines have much in common, each has unique characteristics. Success with programming depends not only on access to a well-developed and tested computing system but also on the user's familiarity with the characteristics and capabilities of the system he is using.

In this chapter we first present a sample model and then describe the use, preparation of data, and interpretation of results for the LP procedure in the Mathematical Programming System/360 (MPS/360) and/or MPSX. This system has been developed and tested by others. You will only be adding data cards and making adjustments in the type and number of control cards used.

WINTERSET EXAMPLE

A simple programming model is developed on the following pages.

The reader should familiarize himself with the structure of the model and its coefficients before proceeding further, since the model is the basis for a series of illustrations in this and later chapters on data preparation, interpretation of output, and special features.

1. Objective:
   To maximize income over variable costs within the restraints imposed.
2. Restraints:
   R01 = labor: 2,400 hours
   R02 = capital: $13,000
   R03 = Winterset silt loam: 150 acres
   R04 = Shelby loam: 18 acres
3. Transfer rows:
   The model as structured requires four transfer rows. They have been labeled as follows:
   R05 = corn transfer. The transfer unit is one bushel.
   R06 = standing meadow transfer. The transfer unit is tons of unharvested hay.
   R07 = hay transfer. The transfer unit is tons.
   R08 = oats transfer. The transfer unit is bushels.
4. Note that corn can be sold for $1.20 per bushel but that meadow (hay or pasture) cannot be sold.
5. Activities:

P01 = a continuous corn growing and harvesting activity on Winterset silt loam. The unit of activity is one acre.
Expected yield: 90 bushels per acre
Variable costs: $30 per acre
Labor: 5 hours per acre
Capital: $20 per acre

P02 = a CCOM activity on Winterset silt loam. The activity unit is four acres including two of corn growing and harvesting, one of oats growing and harvesting, and one of meadow growing.

	Corn	Corn	Oats	Meadow
Expected yield per acre:	90	86	56	3.2
Variable costs per acre:	$28	$30	$14	$9

Labor: 4 hours per acre
Capital: $16 per acre

P03 = a CCOM activity on Shelby loam. The activity unit is four acres including two of corn growing and harvesting, one of oats growing and harvesting, and one of meadow growing.

	Corn	Corn	Oats	Meadow
Expected yield per acre:	64	60	46	1.8
Variable costs per acre:	$29	$33	$16	$11

Labor: 4 hours per acre
Capital: $18 per acre

P04 = a COMM activity on Shelby loam. The activity unit is four acres.

	Corn	Oats	Meadow	Meadow
Expected yield per acre:	66	46	1.8	1.8
Variable costs per acre:	$26	$16	$11	$5

Labor: 3 hours per acre
Capital: $15 per acre

P05 = raising beef calves to be sold in the fall. The activity unit is one cow.
90% calf crop
82% of calves sold (18% saved as replacements)
16% of cows culled and sold
Gross return per unit of activity:

(450 lb. calf) x (.90) x (.82) x ($24.69 per cwt)	= $ 82
(1,000 lb. cow) x (.16) x ($13 per cwt)	= $ 21
Total	$103

Feed requirements per unit of activity:
5 bushels of corn
2 tons of hay
4 tons of pasture equivalent
Variable costs: $24 per unit of activity *(Feed is not included)*
Labor: 20 hours per cow
Capital: $142 per cow

P06 = buying, feeding, and selling yearling steers. The unit of activity is one yearling steer.
Gross return per unit of activity:
1,150 lb. @ $23 per cwt = $265
Original cost of steer: 650 lb. @ $23 per cwt = $150
Feed requirements per unit of activity:
60 bushels of corn
1 ton of hay
1 ton of pasture equivalent

Variable costs: \$23 per steer
Labor: 15 hours per steer
Capital: \$160 per steer

P07 = raising and selling hogs on a two-litter system. The unit of activity is one sow and two litters.
7.3 pigs weaned per litter
1.0 pig per unit of activity retained as a replacement gilt and sold later as a sow
0.2 pig death loss per unit of activity
Gross return per unit of activity:

13.4 hogs @ 220 lb. each @ \$18 per cwt	=	\$530
1 sow @ 400 lb. @ \$15 per cwt	=	\$ 56
Total		\$586

Feed requirements per unit of activity:
210 bushels of corn
1 ton of pasture equivalent
Variable costs: \$140 per unit of activity
Labor: 36 hours per unit of activity
Capital: \$90 per unit of activity

P08 = selling corn. The unit of activity is one bushel and the price is \$1.20 per bushel.

P09 = harvesting hay. The unit of activity is one ton.
Variable costs: \$4 per ton
Labor: one hour per ton
Capital: \$3 per ton

P10 = corn buying. The unit of activity is one bushel and the price is \$1.25 per bushel.
Capital: \$.40 per unit of activity

P11 = selling oats. The unit of activity is one bushel and the price is \$.67 per bushel.

The information presented above is summarized in tabular form on page 84.

DATA PREPARATION FOR MPS/360 AND/OR MPSX LINEAR PROGRAMMING

Standard Procedures

1. The first card in the data deck will always be a name card. The letters NAME are entered in columns 1 through 4 of the first row of the data sheet (Fig. 4.1). You may assign any name you wish to your problem that can be fitted into columns 15–22 on the name card, except that the name given must correspond to the name you enter on the MØVE XDATA card in the control deck. The control deck listings used in the following examples contain the name ECØN430.
2. The second step in data preparation is to identify the row

WINTERSET EXAMPLE

Row Type		B	P01	P02	P03	P04	P05	P06	P07	P08	P09	P10	P11
N	C		—30	—81	—89	—58	79	92	446	1.20	—4	—1.25	.67
L	R01	2,400	5	16	16	12	20	15	36		1		
L	R02	13,000	20	64	72	60	142	160	90		3	.4	
L	R03	150	1	4									
L	R04	18			4	4							
L	R05	0	—90	—176	—124	—66	5	60	210	1		—1	
L	R06	0		—3.2	—1.8	—3.6	4	1	1		1		
L	R07	0					2	1			—1		
L	R08			—56	—46	—46							1

MPS/360 Linear Programming

80 COLUMN DATA SHEET

PROGRAM | JOB NO. | BY | DATE

1 Type 4	5 Name 12	15 Name 22	25 Coefficients 36	40 Name 47	50 Coefficients 61	70	80
NAME		ECØN430					
RØWS							
N	C						
L	R01						
L	R02						
L	R03						
L	R04						
L	R05						
L	R06						
L	R07						
L	R08						
CØLUMNS							
	P01	C	-30.	R01	5.		
	P01	R02	20.	R03	1.		
	P01	R05	-90.				
	P02	C	-81.	R01	16.		
	P02	R02	64.	R03	4.		
	P02	R05	-176.	R06	-3.2		
	P02	R08	-56.				
	P03	C	-89.	R01	16.		
	P03	R02	72.	R04	4.		
	P03	R05	-124.	R06	-1.8		
	P03	R08	-46.				
	P04	C	-58.	R01	12.		

FIG. 4.1. The Winterset Model prepared for processing: Data Sheet I.

names within the model. The second data card must contain the letters RØWS in the first four columns.

3. The objective function should be labeled C in column 5 and should be the first row named. The other rows can have any name of not more than 8 characters entered in columns 5–12. Use of the traditional R01, R02, . . . RX labeling system is recommended.
4. The type of restraint must be specified for each row named in the row identification section. Column 3 is used for this purpose. The row type is specified according to the following code:

N = objective function or nonrestrictive row
G = minimum restraint (greater than or equal to)
L = maximum restraint (less than or equal to)
E = equality

5. After the rows have been identified and typed, an indicator card labeled "columns" follows. The word CØLUMNS is entered in the first 7 columns (see Fig. 4.1).
6. You are then ready to begin entering the coefficients. You first enter the column name in columns 5–12, the row name in columns 15–22, and the coefficient value in columns 25–36. Commas which appear in models in numbers 1,000 or greater for ease of reading should not be entered on the data sheets.
7. You do not begin with the B column even though it may appear first in your model. B column entries must follow the other coefficients in the model.
8. The first entry will normally be the coefficient at the intersection of P01 (the column) and C (the row) with the coefficient value placed in columns 25–36. Coefficients with zero values are not entered on the data sheet.
9. All column and row names must be left-justified (begin in the leftmost column) in their fields and may have no embedded blanks.
10. Note that two coefficients can be entered on the same card (or row of the data sheet). The name of the column is not repeated for the second coefficient on a card.
11. There are 12 columns in each of the two fields (i.e., columns 25–36 and 50–61) reserved for coefficients. The decimal point has been fixed to permit seven places to the left and four places to the right of the decimal.
12. Specification of a positive sign for coefficients is optional. If none is specified a plus (+) sign is implied. Minus (—) signs must be specified and should be placed in the field preceding the number.
13. The B column is the last column to be recorded in the standard data deck. It is preceded by a card with RHS in columns 1–3 (see Fig. 4.2).
14. The data deck terminates with a card containing ENDATA in columns 1–6.
15. The control cards required for a standard run are listed on page 88. A standard procedure is one wherein (1) the objective function is maximized, (2) only one optimization results, and (3) the bounding option is not utilized.
16. In some cases it may be necessary or desirable to alter the name given to your model from that shown in the sample deck. The name assigned may contain one to eight letters, numbers, and/or periods in columns 15–22. If you change

MPS/360 Linear Programming
80 COLUMN DATA SHEET

PROGRAM			JOB NO.	BY		DATE

1 Type 4	5 Name 12	15 Name 22	25 Coefficients 36	40 Name 47	50 Coefficients 61	70	80
	P04	R02	60.	R04	4.		
	P04	R05	-66.	R06	-3.6		
	P04	R08	-46.		.		
	P05	C	79.	R01	20.		
	P05	R02	142.	R05	5.		
	P05	R06	4.	R07	2.		
	P06	C	92.	R01	15.		
	P06	R02	160.	R05	60.		
	P06	R06	1.	R07	1.		
	P07	C	446.	R01	36.		
	P07	R02	90.	R05	210.		
	P07	R06	1.		.		
	P08	C	1.20	R05	1.		
	P09	C	-4.	R01	1.		
	P09	R02	3.	R06	1.		
	P09	R07	-1.		.		
	P10	C	-1.25	R02	.4		
	P10	R05	-1.		.		
	P11	C	.67	R08	1.		
RHS			.		.		
	B	R01	2400.	R02	13000.		
	B	R03	150.	R04	18.		
ENDATA			.		.		
			.		.		

FIG. 4.2. The Winterset Model prepared for processing: Data Sheet II.

the card NAME ECØN430 you must also change the MØVE (XDATA, 'ECØN430') card in exactly the same way.

17. The NAME and the ENDATA cards are the first and last cards of the data deck respectively. The RØWS, CØLUMNS, and RHS sections must follow the NAME card and precede the ENDATA card.
18. The reader should note that programming applications involve three groups of cards: (1) systems cards, (2) control cards, and (3) the data deck. In using the MPS/360 and MPSX systems, most modifications in the computing routine are effected through manipulation of the control program cards.
19. No effort has been made to specify the job control language cards that are needed to compile and execute the control program, because they undergo frequent revisions causing particular cards to become obsolete and also will vary from one computer installation to another.

STANDARD PROCEDURE

Job Control Language and Control Program Cards

```
JOB CONTROL LANGUAGE CARDS

               PROGRAM
               INITIALZ
               MOVE(XDATA,'ECON430')
               MOVE(XPBNAME,'PBFILE')
               MVADR(XMAJERR,UNB)
               MVADR(XDONFS,NOF)
               CONVERT
               SETUP('MAX')
               MOVE(XRHS,'B')
               MOVE(XOBJ,'C')
               PRIMAL
               SOLUTION
               CHECK
               EXIT
NOF            TRACE
UNB            EXIT
               PEND
/*

JOB CONTROL LANGUAGE CARDS

NAME                ECON430

ENDATA
/*
```

Bounding Activities

The MPS/360 and MPSX systems provide a bounding feature for restraining the level of any activity. The reader should recall that maximum, minimum, or equality restraints can be built into the model as demonstrated in Chapter 3. Thus knowledge of bounding procedures is not essential to programming applications. However, it can result in a much less complicated model, particularly in situations where large numbers of activity restraints are contemplated.

The steps using this feature are as follows:

1. The bounds section of the data deck is preceded by the letters BØUNDS in columns 1–6.

2. Because it is possible to obtain multiple solutions in the same computer run based upon different sets of bounds, you must give a name to the bound rows. Use the names BND1, BND2, . . . BNDX to distinguish different bound sets. Even though you are using only one set of bounds, the row name BND1 should be given to the bound row.
3. You may specify upper (maximum) bounds, lower (minimum) bounds, or fixed (equality) bounds or combinations thereof on activities.

   The type of bound intended is entered in columns 2 and 3 according to the following symbols:

   UP = upper bound or maximum
   LO = lower bound or minimum
   FX = fixed bound or equality

MPS/360 Linear Programming
80 COLUMN DATA SHEET

PROGRAM			JOB NO.	BY		DATE	
1 Type 4	5 Name 12	15 Name 22	25 Coefficients 36	40 Name 47	50 Coefficients 61	70	80
			DATA FORMAT FOR BOUNDING OPTION				
	The BOUNDS section follows immediately the RHS section in the data deck. (Note the entry on the SETUP card						
	BND1 which signals the presence of the bounds coefficients.)						
BØUNDS							
LØ	BND1	P05	15.	Columns 40 through 61 are never used			
UP	BND1	P07	30.	in defining BOUNDS coefficients.			
			DATA FORMAT FOR MULTIPLE B COLUMNS				
	These data are part of the RHS section and follow immediately the B coefficients.						
	B2	R01	2200.	R02	13000.		
	B2	R03	210.	R04	18.		
	B3	R01	2200.	R02	16000.		
	B3	R03	260.	R04	12.		
			DATA FORMAT FOR MULTIPLE C ROWS				
	Note that additional C rows must be named in the Rows section.						
NAME		ECØN430					
RØWS							
N	C						
N	C2						
REMAINING ROW NAMES HERE							
CØLUMNS							
	P01	C	-30.	C2	-40.		

FIG. 4.3. Data sheet format showing optional features prepared for processing.

BOUNDS OPTION

Job Control Language and Control Program Cards

```
JOB CONTROL LANGUAGE CARDS


              PROGRAM
              INITIALZ
              MOVE(XDATA,'ECON430')
              MOVE(XPBNAME,'PBFILE')
              MVADR(XMAJERR,UNB)
              MVADR(XDONFS,NOF)
              CONVERT
              SETUP('MAX','BOUNDS','BND1')
              MOVE(XRHS,'B')
              MOVE(XOBJ,'C')
              PRIMAL
              SOLUTION
              EXIT
NOF           TRACE
UNB           EXIT
              PEND
/*

JOB CONTROL LANGUAGE CARDS


NAME                ECON430


ENDATA
/*
```

4. The bound name BND1 appears in columns 5–8. The activity (column) with which the bound is associated appears in columns 15–22. The level at which the activity is to be bounded appears in columns 25–36.
5. Columns 40–47 and 50–61 are never used in specifying bounds.
6. You may specify both an upper and a lower bound on the same activity.
7. When no bounds are specified, the system assumes a zero lower bound and an upper bound of infinity.
8. To fix a bound at zero the FX bound must be used.

MULTIPLE BOUNDS

Job Control Language and Control Program Cards

```
JOB CONTROL LANGUAGE CARDS


                PROGRAM
                INITIALZ
                MOVE(XDATA,'ECON430')
                MOVE(XPBNAME,'PBFILE')
                MVADR(XMAJERR,UNB)
                MVADR(XDONFS,NOF)
                CONVERT
                SETUP('MAX','BOUNDS','BND1')
                MOVE(XRHS,'B')
                MOVE(XOBJ,'C')
                PRIMAL
                SAVE
                SOLUTION
                SETUP('MAX','BOUNDS','BND2')
                RESTORE
                PRIMAL
                SAVE
                SOLUTION
                SETUP('MAX','BOUNDS','BND3')
                RESTORE
                PRIMAL
                SOLUTION
                EXIT
NOF             TRACE
UNB             EXIT
                PEND

JOB CONTROL LANGUAGE CARDS

NAME                ECON430


ENDATA
/*
```

Multiple B Columns

1. The card order for a model containing two B columns is shown on the card printout labeled "Multiple B (B, B2)" and for three B columns labeled "Multiple B (B, B2, B3)." The control deck can be extended to include any number of B

columns by reproducing (with the appropriate B column label for each B column contemplated) the control cards from MØVE through SØLUTIØN.

2. The B column names and coefficients are always entered on the data sheet under the RHS section. An illustration of the data format for multiple B columns is shown on an accompanying sheet.
3. The names given B columns on the data sheet must correspond exactly to those contained on the control cards. Because the convention of naming B columns B, B2, B3, . . . BX has been followed in preparing control cards, you should follow this system in labeling the B columns in the RHS section of the data deck.

Multiple C Rows

1. The convention has been followed in preparing control cards of labeling the original objective function C and subsequent functions C2, C3, . . . CX.
2. Each C row contained in the model must be labeled in the RØWS section in the data deck. All C rows must be preceded by the letter N in column 3. The name of the row (C, C2, C3, etc.) appears in columns 5 and 6, left-justified.
3. In entering the coefficients C row data are treated the same as data from other rows. The column name is given in columns 5–12, the row name in columns 15–22, and the coefficient in columns 25–36. A second row (e.g., C2) may be named in columns 4–47 and its coefficient in columns 50–61.
4. The control cards are modified from a standard deck as specified in the card printout labeled "Multiple C Rows."

Combination Multiple B Columns and Multiple C Rows

1. Any combination of B columns and C rows may be included in one model and a single computer run.
2. The same rules apply in preparing the data deck as have been given in the sections on multiple B columns and multiple C rows.
3. The control card order is reproduced on the following pages.
4. The major source of difficulty in preparing multiple runs is failure to maintain consistency in labeling the multiple rows

and columns. The convention shown on the control cards should be adhered to carefully in naming rows and labeling coefficients in the data deck.

MULTIPLE B (B, B2)

Job Control Language and Control Program Cards

```
JOB CONTROL LANGUAGE CARDS


                PROGRAM
                INITIALZ
                MOVE(XDATA,'ECON430')
                MOVE(XPBNAME,'PBFILE')
                MVADR(XMAJERR,UNB)
                MVADR(XDONFS,NOF)
                CONVERT
                SETUP('MAX')
                MOVE(XRHS,'B')
                MOVE(XOBJ,'C')
                PRIMAL
                SAVE
                SOLUTION
                MOVE(XRHS,'B2')
                RESTORE
                PRIMAL
                SOLUTION
                EXIT
NOF             TRACE
UNB             EXIT
/*

JOB CONTROL LANGUAGE CARDS


NAME            ECON430


ENDATA
/*
```

MULTIPLE B (B, B2, B3)

Job Control Language and Control Program Cards

```
JOB CONTROL LANGUAGE CARDS


                PROGRAM
                INITIALZ
                MOVE(XDATA,'ECON430')
                MOVE(XPBNAME,'PBFILE')
                MVADR(XMAJERR,UNB)
                MVADR(XDONFS,NOF)
                CONVERT
                SETUP('MAX')
                MOVE(XRHS,'B')
                MOVE(XOBJ,'C')
                PRIMAL
                SAVE
                SOLUTION
                MOVE(XRHS,'B2')
                RESTORE
                PRIMAL
                SAVE
                SOLUTION
                MOVE(XRHS,'B3')
                RESTORE
                PRIMAL
                SOLUTION
                EXIT
NOF             TRACE
UNB             EXIT
                PEND
/*

JOB CONTROL LANGUAGE CARDS


NAME            ECON430


ENDATA
/*
```

Interpretation of Output

One page of output from a standard solution for the Winterset example is reproduced in Figure 4.4. At this writing the output format for MPS/360 has not been modified to facilitate interpretation in farm planning applications. The terminology used does not follow the

MULTIPLE C (C, C2)

Job Control Language and Control Program Cards

```
JOB CONTROL LANGUAGE CARDS


           PROGRAM
           INITIALZ
           MOVE(XDATA,'ECON430')
           MOVE(XPBNAME,'PBFILE')
           CONVERT
           SETUP('MAX')
           MOVE(XRHS,'B')
           MOVE(XOBJ,'C')
           PRIMAL
           SAVE
           SOLUTION
           MOVE(XOBJ,'C2')
           RESTORE
           PRIMAL
           SOLUTION
           EXIT
NOF        TRACE
UNB        EXIT
/*         PEND


JOB CONTROL LANGUAGE CARDS


NAME            ECON430


ENDATA
/*
```

pattern of our previous discussion; it is presented in a cost minimization framework, whereas our interest has been in maximizing income. The strange terminology notwithstanding, the results are fully applicable to farm planning applications once you learn to properly interpret the output.

Our interest centers on Sections 1 and 2 of the output. Section 1 gives the value of the program in the C row of the column labeled ACTIVITY. The remaining entries in this column indicate how much of the original B column value was used in the production

MULTIPLE B – MULTIPLE C (B, B2, B3; C, C2)

Job Control Language and Control Program Cards

```
JOB CONTROL LANGUAGE CARDS

              PROGRAM
              INITIALZ
              MOVE(XDATA,'ECON430')
              MOVE(XPBNAME,'PBFILE')
              MVADR(XMAJERR,UNB)
              MVADR(XDONFS,NOF)
              CONVERT
              SETUP('MAX')
              MOVE(XRHS,'B')
              MOVE(XOBJ,'C')
              PRIMAL
              SAVE
              SOLUTION
              MOVE(XRHS,'B2')
              RESTORE
              PRIMAL
              SAVE
              SOLUTION
              MOVE(XRHS,'B3') .
              RESTORE
              PRIMAL
              SAVE
              SOLUTION
              MOVE(XRHS,'B')
              MOVE(XOBJ,'C2')
              RESTORE
              PRIMAL
              SAVE
              SOLUTION
              MOVE(XRHS,'B2')
              RESTORE
              PRIMAL
              SAVE
              SOLUTION
              MOVE(XRHS,'B3')
              RESTORE
              PRIMAL
              SOLUTION
              EXIT
NOF           TRACE
UNB           EXIT
              PEND
/*            PEND

JOB CONTROL LANGUAGE CARDS

NAME          ECON430

ENDATA
/*
```

SECTION 1 - ROWS

NUMBER	...ROW..	AT	...ACTIVITY...	SLACK ACTIVITY	..LOWER LIMIT.	..UPPER LIMIT.	.DUAL ACTIVITY
1	C	BS	19416.99302	19416.99302-	NONE	NONE	1.00000
2	R01	UL	2400.00000	.	NONE	2400.00000	4.28525-
3	R02	BS	7205.35252	5794.64748	NONE	13000.00000	.
4	R03	UL	150.00000	.	NONE	150.00000	56.57374-
5	R04	UL	18.00000	.	NONE	18.00000	35.90709-
6	R05	UL	.	.	NONE	.	1.20000-
7	R06	UL	.	.	NONE	.	39.73094-
8	R07	BS	.	.	NONE	.	.
9	R08	UL	.	.	NONE	.	.67000-

SECTION 2 - COLUMNS

NUMBER	.COLUMN.	AT	...ACTIVITY...	..INPUT COST..	..LOWER LIMIT.	..UPPER LIMIT.	.REDUCED COST.
10	P01	BS	113.56835	30.00000-	.	NONE	.
11	P02	BS	9.10791	81.00000-	.	NONE	.
12	P03	LL	.	89.00000-	.	NONE	50.05669-
13	P04	BS	4.50000	58.00000-	.	NONE	.
14	P05	LL	.	79.00000	.	NONE	171.62878-
15	P06	LL	.	92.00000	.	NONE	84.00971-
16	P07	BS	45.34532	446.00000	.	NONE	.
17	P08	BS	2598.62590	1.20000	.	NONE	.
18	P09	LL	.	4.00000-	.	NONE	48.01619-
19	P10	LL	.	1.25000-	.	NONE	.05000-
20	P11	BS	717.04317	.67000	.	NONE	.

CHECK

ROW NAME	UPPER LIMIT	LOWER LIMIT	ROW ERROR
C	NONE	NONE	.
R01	2400.0000000	NONE	.
R02	13000.0000000	NONE	.
R03	150.0000000	NONE	.
R04	18.0000000	NONE	.
R05	.	NONE	.
R06	.	NONE	.
R07	.	NONE	.
R08	.	NONE	.

FIG. 4.4. Computer output report for Winterset Example.

process. In this example all of R01, R03, and R04 were used. The original B column entry for R02 was $13,000. Of this amount $7205.35 was put to use in the program. The values for R05, R06, R07, and R08 (all transfer rows) are zero.

The shadow prices for the disposal or slack activities are printed in the column labeled DUAL ACTIVITY. You should ignore the C row value shown in this column. The remaining values specify the change in the value of the program which would result from one less unit of restraint (or resource) in the original B column entry. In the Winterset example all values are followed by a minus sign, indicating (as one should expect) that tightening the resource restraints would lessen the value of the program. R02 (capital) is in excess supply and therefore has a shadow price of zero. Zeros in the output appear as blanks except for the decimal point.

Section 2 provides information on the real activities in the solution. Activity levels are printed out under a column labeled ACTIVITY. The level of activity appears in the output report in the units in which the activity was defined in constructing the model.

The column INPUT CØST only repeats the net prices assigned in the original model. Hence they have no significance in interpreting the output report except as a means of checking to see that they correspond to the values originally intended.

The lower and upper limit columns will contain meaningful entries only when the original model includes provisions for bounding the activities. In the latter case any bounds imposed are printed out as a reminder.

The reduced cost column shows the income penalties associated with forcing into the solution one unit of activity. P03, for example, is not in the solution (i.e., its value is zero). Under the reduced cost column we find that forcing one unit of P03 into the solution would decrease the value of the program by $50.06.

The output report also contains a section labeled CHECK. The column on row error is of principal interest in this section. The entries in this column will be blank (except for decimal points) indicating zeros. The CHECK cards (which report the check sections) have been removed from all control programs listed except the standard program. The MPS/360 or MPSX algorithms check for row errors even if the control card CHECK is not present. The steps performed in correcting a computational error (should an element in the row error column be nonzero) is not dependent upon the CHECK section being reported.

TABLE 4.1: **Comparison of Computer Output with Simplex Tableaus**

Computer Output	Simplex Tableaus
Section 1.	
Activity	difference between Section 1 resource levels (B column coefficients) and final section slack activity levels
Slack activity	slack activity levels in the final section (B column if basic)
Lower limit	level specified for any G or E restraint in Section 1
Upper limit	level specified for any E or L restraint in Section 1
Dual activity	Z-C row elements for disposal activities in the final section (shadow prices on resources)
Section 2.	
Activity	level of real activity in the B column of final section
Input cost	C row elements
Reduced cost	Z-C row elements for real activities in the final section (shadow prices for real activities)

The relevant information received in the computer printout is equivalent to the items derived by the iterative process, except for the signs attached to the shadow prices. Table 4.1 indicates where or how the items found in the computer printout may be found or derived from the simplex tableaus discussed in Chapter 2.

CHAPTER FIVE

FORMING PRICE AND PRODUCTION COEFFICIENTS

Depending upon the structure of the model, the degree to which per unit input or product prices are expressed explicitly varies greatly. For example, forming the C row coefficient for a swine raising and selling activity may involve aggregating the total pounds of pork to be sold per unit of activity and subtracting a collection of variable costs. In this case several input and product prices may be embedded in a single coefficient. At the other extreme the price per bushel appears unmodified in a corn buying or selling activity. By structuring combinations of transfer rows and buying or selling activities into the model, the price of any input or product can be made explicit in the objective function. Thus in the case of swine production one could have transfer rows for butcher hogs marketed in July, August, September, and February and for packing sows in August with corresponding selling activities. He could also include purchasing activities and appropriate transfer rows and coefficients for all types of feed and services purchased.

There are several advantages to disaggregating activities and hence their C row coefficients to a greater degree than has been done in the models thus far presented. For one thing, errors are likely to be reduced, and those that do occur are more easily detected. Additionally, it is not unusual to want to change price coefficients either in a preplanned manner such as with the multiple C row or parametric routine or simply through a series of adjustments, each involving a single optimization. In either case the process of altering the C row coefficient is greatly facilitated if buying and selling activities are used and the coefficient is stated in terms of price per unit for a single product or input.

The disadvantage to a model with provisions for large numbers of buying, selling, and transfer rows is that its size may be greatly increased. However, this is an instance where one should not equate size with the amount of effort involved in constructing the model.

Once one is thoroughly confident of his ability to deal with transfer rows and buying and selling activities, there is little more effort involved in a model with highly disaggregated activities than in one where many functions are combined under a single activity. There will likely be fewer errors because arithmetic calculations will be performed by machine instead of by hand. Larger models may involve additional machine time, but the difference when working with an efficient computing routine and system is not likely to be significant.

PRICES

Regardless of the manner in which coefficients are expressed, they must be built upon the planner's expectation of prices. This is not a condition peculiar to linear programming. All planning requires formulation of price expectations either implicitly or explicitly. As in the case of all planning methods, the results from a linear programming analysis are meaningless if price projections have not been well conceived.

A first step in the process of forming price coefficients for any planning operation is to relate the expectations to the length of the horizon visualized for the plan. If it is strictly a one-year plan, then price relationships should reflect this fact. On the other hand, if one is laying out a long-run strategic plan to be adjusted marginally from year to year, then he must take the long-run view of prices.

Price Expectation Models

Studies have demonstrated that farm producers and others use a variety of models in formulating price expectations. One simple method is to project present prices indefinitely into the future. Students quite generally exhibit a marked preference for this model, probably because of the ease with which it can be applied. Another model often used, particularly in the case of product prices, is to project some long-term average relationship into the future. This has the advantage where long-term plans are being prepared, because price relationships based on several years of history exhibit more stability over time than those formed from a single point in time. Indeed, the latter may be highly misleading if one is unlucky enough to base his outlook on a time when price relationships were highly distorted. Still another possibility is to form long-term prices on the basis of past price relationships adjusted for trend or to take account of some clearly discernible force which can be expected to alter price relationships in the future.

Cyclical and Seasonal Price Patterns

Several product prices have demonstrated a marked tendency to move in cyclical patterns. Hogs are the most striking example. The causes of the cycles seem to be related almost completely to the collective response of producers to expected price relationships. A few producers are successful in synchronizing their level of production through short-run adjustments in output to take advantage of cyclical price changes. However, most producers are not. Hence, short-term adjustments in price coefficients to reflect expected cyclical changes are not likely to be accurate. A planning period of sufficient length to permit one to average expected prices over one or more cycles is a more solid foundation for planning.

Seasonal price variations, however, are a different story. Such patterns recur with sufficient regularity that they should be taken into account in forming price coefficients. The most straightforward manner in which to reflect seasonality is to view the same product sold in different months as distinct activities. Hence, hog selling in September is a separate activity from hog selling in December. In addition, the difference in the date at which hogs are ready for market should be reflected in separate producing activities, because demands on resources may vary both in their timing and magnitude.

Price coefficients for grain products should also reflect the date at which one expects to market the grain. Typically grain prices are lowest after harvest and rise gradually until shortly before the new harvest. When some time elapses between harvest and sale, the producer is not only producing and selling grain but is also storing it. Since storage on the farm involves mostly fixed costs which are reflected in the fixed cost paying activity, it is reasonable to incorporate the return for storage in the selling price. On the other hand, where no farm storage occurs because the crop is expected to be sold immediately upon harvest, the price coefficient should reflect this fact. Likewise, where storage services are hired, the planner should either include a storage activity or simply deduct the storage charge from the anticipated price. The important point to bear in mind in forming coefficients is that price expectations related to the month of expected sale are more meaningful than an average price for the year.

PRICING OF INPUTS

For purposes of discussion we may divide inputs into three categories: those that appear in the model in the form of B column coefficients, those that are deducted in forming price coefficients in a producing activity, and those that are brought into the model by purchasing activities and corresponding transfer rows.

The B column quantities never appear as a charge against the value of the program. Instead they are imputed a share of the value of the program. Stated differently, the value of the program is a return for the inputs which appear in the B column of the original model. If the operator's labor, land, and capital are introduced into the model as B column quantities and appropriate fixed costs have not been charged, the value of the program is a return for the services of these resources. This is little different from the interpretation one gives to the conventional net farm income figure. Although this is the traditional method of approaching the resources supplied by the operator, the planner is not restricted to treating operator-supplied resources in this manner. By relatively simple alterations in the structure of the model, he may place an opportunity cost on the operator's resources and arrange (with appropriate restraints) for the business to "purchase the services of the operator from himself." In this instance the value of the program approaches the concept of management return as conventionally used in farm accounting.

In the case of all non-B column inputs, the planner is required to form price expectations. These may be as critical to the usefulness of the plan as the coefficients for product prices. The fact that some price coefficients for inputs may not appear explicitly in the model does not lessen their impact on the optimum plan. Importance of price coefficients for inputs is clear in the case of purchased feeder livestock or grain. However, other inputs such as seed, fertilizer, herbicides, and commercially prepared livestock feeds are purchased in such volume in modern agriculture that their price coefficients also are of great importance. Most of the same problems exist in forming coefficients for livestock and feed purchased as exist for product prices. They are, after all, the other side of the same coin.

With the exception of the mixed livestock feeds which contain a heavy proportion of farm-raised ingredients, prices of inputs exhibit more stability than do farm product prices. Hence present prices or prices during the last year are not likely to lead to large distortions when used as the basis for formulating price coefficients.

Quality of Products and Purchased Inputs

Price coefficients should reflect the quality of product or input expected to be produced or purchased. In the case of livestock, particularly beef cattle, the differences are obvious and are not often overlooked in the planning process. Other less obvious but important quality differences should be taken into account. In the case of corn, for example, the percentage of moisture in the product may be the source of important quality variation. If price expectations are based on corn which has been dried artificially, a drying activity should be included along with an appropriate C row coefficient. Or if one anticipates selling corn immediately after harvest with a moisture content which brings the corn below grade, the anticipated price discount should be made when forming the C row coefficient.

Prices received for different grades of hogs typically do not vary greatly for butcher hogs of the same weight, but weight is an important quality component. Therefore, the weight at which the planner expects to market hogs should be taken into account in forming price expectations. Packing sows and butcher hogs vary so widely in price that it is best to treat them as separate products and form separate selling activities for each.

The pricing structure for dairy products is often complex; one must know well the price structure within the market involved to form meaningful price coefficients. There will probably be markets for more than one quality of product. Additionally, many markets have provisions for temporarily assigning a lower use and hence a lower price for a portion of all Grade A milk produced. Models which do not take into account the peculiarities of the market structure where the farm in question is located are not likely to be of much value where dairying is a potentially competitive activity.

MARKETING COSTS

Moving products from farm to market and through the market involves costs in the form of transportation, commission, and service charges. They may be charged directly to the seller or indirectly through a lower price. In any case, the appropriate price to use in farm planning for both products and inputs is the price at the farm gate. The difference may not be great in the case of local markets, although in the case of milk the difference can be substantial. Adjustments should always be made for products sold or inputs purchased at distant markets and for milk even when sold locally. In other cases, adjustment for marketing costs is not imperative, but the adjustments can be made with such little effort that C row coefficients should be formed to reflect farm gate prices.

PRODUCTION COEFFICIENTS

All planning implies knowledge of the relevant input-output coefficients. Although construction of programming models requires almost unlimited quantities of precise data, no more are required than by other methods if one seeks the same degree of effectiveness. The planner cannot hope to achieve perfection in formulating coefficients because he is dealing with the uncertainties of weather, pests and diseases, and human performance; but there are measures he can take that will improve greatly the quality of coefficients.

Sources of Data

Ideally, production coefficients would be derived from carefully kept input-output records on the farm under study. But this is rarely possible. Many planning models will seek to test activities which have never been carried out on the farm; hence no record is possible. Furthermore, the detailed accounting records required to develop planning coefficients are too costly in terms of time and effort to be justified on units of the size typical in much of today's agriculture. Thus the assumption that all would be easy if only adequate records were available is a useless speculation. Such records simply are not available for farm planning, nor will they be in the foreseeable future.

Most production coefficients must be built on knowledge transferred from another situation and adapted as best one can to the business under study. Two likely sources of data which may be transferred are (1) experimental data and (2) cost accounting data. In the former case, the data are a by-product of a project conducted for other purposes. Cost accounting data come from detailed accounts kept under farm conditions by an experiment station, a private industry, or a cooperative group specifically for the purpose of supplying information to a large group. In this manner the cost is shared by a large number of farm operators.

Operator Recall

Farm operators usually can recall a surprising collection of information about their businesses if a relatively recent time period is involved. If data are sought concerning events after more than a year has elapsed, confusion among years is often evident.

Information will be more consistent and dependable if it is sought in small pieces. Furthermore, questions should be specific. Questions like How much did your corn yield last year? or How much fertilizer did you use? are unlikely to elicit a useful response. A more effective approach is to begin questions concerning crop history by asking the operator to prepare with you a map of the farm with the area and crop of each field identified. Then highly specific questions can be formulated: (1) How many loads of corn were harvested from field A? (2) How many bushels were in a load? (3) What was the moisture content of the corn from this field at harvest? In the case of fertilizer, one should be very specific about the treatment of each field by asking questions such as: (1) How many pounds of plow-down fertilizer were used on field A in the fall? (2) What was the analysis? (3) How many pounds of starter fertilizer were used on field A? (4) What was the analysis of this fertilizer? After all possible methods of fertilizer use have been proved for each field, it is likely that you will have a dependable record of fertilizer use.

The same step-by-step technique is also appropriate for other types of information. Consider the problem of estimating the amount of capital available to the farm business. If the planner proceeds to inventory the livestock type by type and lot by lot and the feed supply by type of grain in each storage structure, he can estimate more accurately the amount of capital available. In addition, he will want to probe sources of credit available to the operator and his willingness to use them.

There is little to be gained by asking the operator what he thinks would happen in the case of activities with which he has had no experience. For example, one should not expect to gather useful information concerning the yield of sorghum silage, the response of pastures to fertilization, or the effect of herbicides on corn yields where the operator has had no experience with the crop or practice. Experimental data properly adjusted are a much more dependable source for this type of information.

Level of Management and Production Coefficients

Production coefficients are the principal means through which variations in the management level can be reflected in planning models. For reasons not easy to define, two operators may achieve quite different results from what appear to be similar activities. One may perform the same field operations, plant the same variety of seed, and use the same mix and level of fertilizer as his neighbor but harvest less corn per acre. The different result may be due to a slight difference in timing or the thoroughness with which operations are performed. The latter are inputs which do not lend themselves to being included explicitly in the planning model. Where information on which to make judgments is available, preferably based on past performance, appropriate adjustment should be made in yield coefficients for crop activities.

Differences in management level of livestock activities may be reflected in the feed conversion rate, the mortality rate, the quality of the product, and the length of the production period. All these manifestations of the level of management can and should be expressed in one way or another by the coefficients specified in the planning model. The length of the production period may be the least straightforward but may be reflected in part through the feed coefficients. The longer production period typically associated with poor livestock performance should also be expressed through higher labor coefficients.

Labor Coefficients

In constructing most models one can expect to encounter the greatest difficulty in forming labor coefficients. The input of labor required per unit of output exhibits wide variability from one farm to another. The pace at which the operator and his help work, the operator's capacity to organize his work, the level of mechanization, and the type and topography of the land on the farm are all important sources of variability. Additionally, there is no academic discipline within the experiment stations which has a primary interest in farm labor.

Labor requirements also have a time dimension. It may be as important to the outcome of the planning process to know *when* labor will be required as to know *how much* will be needed.

There are several precautions one should take in adapting labor coefficients from cost accounting data gathered on other farms.

1. Labor coefficients should be adjusted to correspond to the level of mechanization one is assuming in the model. Where one optimizes separately for different levels of mechanization, labor coefficients should be adjusted to conform to the appropriate level. The size and capacity of power and machinery units are obviously important in forming crop production coefficients. But labor requirements for livestock activities also vary markedly, depending on the degree of mechanization.

2. The operation of a farm business involves labor that is not directly allocable to any enterprise. Examples may be repair of fence and multiple-use buildings, maintenance of the farmstead and water system, weed control on roadsides and fencerows, and a variety of procurement activities such as shopping for new machinery. The existence of such labor requirements should be recognized in a fixed labor use activity.

3. Where activities vary from the typical activity in a way that clearly alters the labor requirement, appropriate adjustments should be made. Production and sale of purebred livestock is a case in point; additional alteration is required in record keeping and reporting, in

separation and identification of animals, and in negotiating sales.

4. In gathering labor coefficients special attention should be given to those time periods where labor requirements are known to peak on the type of farm under study. The problem of realistic restraints for critical operations is discussed in greater depth in Chapter 6.

5. Labor required per unit of output is typically a function of the scale on which an activity is conducted. To be useful, information in respect to labor—especially in the case of livestock activities—must be related to the size of enterprise. When one uses data collected by others to form labor coefficients, he should make note of the level of activity for which the information is relevant. The question of the relationship between the scale of enterprise (level of activity) and labor requirements raises the linearity issue. Although this problem has been referred to previously, understanding it is of such importance that we examine it in more detail below.

THE LINEARITY PROBLEM

The student is unlikely to appreciate fully the complication of the linearity assumption until he has pondered the problem of assigning an appropriate coefficient for labor. The dilemma he encounters is this: The labor required per unit of activity is a function of the level at which the activity is conducted. But the coefficient must be specified before the model is optimized and hence before the level of activity is known. Furthermore, labor is one among several classes of inputs where the quantity required per unit of activity decreases as the level of activity increases (or reciprocally, the quantity of activity increases per unit of labor).

There is no satisfactory escape from this dilemma. One cannot, as in the case of increasing input requirements per unit of activity, dissect the activity into a number of component activities, although inexperienced programmers are often tempted to do so. One example will illustrate the impasse which develops with this procedure. Suppose the activity in question is dairy production and that the anticipated labor requirements may be approximated as follows:

Number of Cows	Average Labor Required per Cow	Added Labor Required per Cow
5	100	. .
10	90	80
15	80	60

We then proceed to define three separate dairy activities as follows:

D_1 = 1–5 cows
D_2 = 6–10 cows
D_3 = 11–15 cows

Activities D_1, D_2, and D_3 are each restrained explicitly to five cows and the appropriate labor coefficient (the added labor required) specified for each. Thus the labor coefficients are $D_1 = 100$, $D_2 = 80$, and $D_3 = 60$; all other coefficients are identical. If we attempt to optimize and labor is restricting, D_3 will always enter before D_1 or D_2, and D_3 and D_2 before D_1, because they have lower labor requirements specified in the model. The plan that results would provide for the 11th to 15th cow but not the 1st through 10th cow. Clearly, such a proposal is meaningless.

If we were confronted with increasing labor requirements, the method would be satisfactory; D_1 would always enter the basis before D_2, because D_2 would require larger quantities of labor per unit of activity.

One might take several steps to mitigate the distortions inherent in the linearity assumptions. In choosing a coefficient one should always select the magnitude that corresponds to the level at which one expects the activity to enter. The appropriate coefficient is the average requirement expected at this level. After optimization, one of four situations might obtain: (1) The activity is in the basis at about the level anticipated. In this instance no adjustments are necessary. (2) The activity is in at a low level even with the favorable labor coefficients where the enterprise is clearly uneconomic. In this case the activity should be dropped from the model. This can be done most easily by assigning it a negative C row coefficient. (3) The activity is not in the basis. Again no adjustment is required. (4) The activity is in the basis at a level higher than anticipated. The appropriate response here is to adjust the labor coefficient downward to correspond to the scale of activity now in the basis and then reoptimize. During this process it will become evident to the planner that once he reaches a point where the activity enters at a level commonly found on farms in the area, the labor requirement per unit of activity does not decline markedly. Consequently, the activity level probably will not increase substantially upon reoptimization.

CHAPTER SIX

SUITABLE FIELD TIME RESTRAINTS

Reference has been made earlier to multiple labor restraints and tractor time restraints. In this chapter we examine in greater depth methods by which the problem of critical labor and machine peaks can be treated realistically.

Anyone familiar with a typical Corn Belt farm recognizes the manpower (and to a lesser extent the tractor power) crunch that occurs in the April-May-June period and the September-October-November harvesting and fall plowing period. Often the scale of operations, the investment in machinery, and the choice of enterprises are influenced greatly by the amount of time suitable for field operations during these critical periods. Development of a realistic farm planning model requires more refined treatment of field time restraints than we have thus far presented.

The task is not easy. The number of hours or days in which land is in condition to work varies with weather, soil type, and adequacy of the drainage system. Because of the key role the weather plays, field time availability fluctuates from one year to another. The farm plan should not be based on an assumption that favorable weather will prevail every year. On the other hand, one should not base his plans on the most unfavorable season one can foresee. The only foundation available for formulations of field time expectations for long-term planning purposes is weather data correlated with records of field operations. Although such data are not plentiful, there are enough to provide (by careful extrapolation) guidelines for forming field restraints. To be more specific, effective planning requires an estimate of the number of hours fields are in condition to work during the periods from (say) April 15 to May 20 and from May 21 to June 30. Ideally, the planner would have available to him a frequency distribution of the number of hours over perhaps a 50-year period. From this he could develop restraints within some acceptable (to him) probability that he could complete his field operations in a timely manner. Many farm operators would be willing to plan within a probability of 80% of successful completions; i.e., they would plan a program they could execute without costly delays four out of five years.

During a crucial period, the pace at which work progresses on field-suitable days will be determined by one or more limiting factors. Intuitively we look upon manhours as the most likely restraint, but often this is not the case. The number of tractors available and their capacity could be limiting, or the number and size of plows could be the bottleneck.

If one element (dryers, wagons, tractors, or pickers) can be identified as clearly limiting during an operation, there is no point in including additional restraints. A method for diagnosing the most limiting element in the power and machinery complex will be presented in the following section. It is not easy to predict whether labor or power and machinery capacity will be the more restricting on a diversified crop and livestock farm. Livestock activities may compete for labor during critical crop periods. In these situations it is well to include both a labor and a power and/or machine restraint.

IDENTIFYING LIMITING MACHINES

The amount of field work accomplished during one time period may be constrained by (1) the number and capacity of one or more machines, (2) the size and number of tractors available, or (3) the amount of operator time available for use in the field. Rather than form a separate set of restraints (and corresponding coefficients) for each field operation, the operation (or type of equipment) that is most limiting can be determined and a restraint formed from it. Activities carried on at the maximum level permitted by the most limiting restraint would not be affected by less binding restraints.

We may use seedbed preparation and planting corn as an example. Assume that we first spread phosphorus and potassium fertilizer, then disk stalks, plow the ground, and finally disk the plowed land twice before planting. Each operation is powered by a tractor and each tractor requires one operator. The most limiting type of equipment is determined as shown in Table 6.1.

Tractor time becomes the most limiting because more hours of tractor time in the field are required than for any other type of equipment involved in preparation and planting. Although tractor time will often be the most limiting, it need not be in every situation. The planter can be limiting during the planting operation, or a self-propelled corn harvester can become the bottleneck during harvest.

TABLE 6.1: Determining Field Time Requirements and Most Limiting Equipment

Operation	Hours per Acre	No. of Machines	Field Time per Acre (hours)
Spreading phosphorus and potassium	.12	1	.12
Disking cornstalks	.17	1	.17
Plowing	.64	2	.32
Double disking	.34	1	.34
Corn planting	.29	1	.29
Tractor Use	1.56	2	.78 to 1.04*

* If operations can be fully overlapped, the elapsed field time required would be .78 hour. Without overlapping the time would be 1.04 hours.

MODEL 6.1: *Restraints Built on Most Limiting Element in Field Operations*

Explanation

The partial model which follows illustrates the application of field time restraints in preparing land and planting corn. The magnitude of the B column coefficients in R03 is a function of the weather and to a lesser degree of soil type. Tractor field time is the product of (1) the number of work days in the period, (2) the percentage of days suitable for field work, (3) the number of operators (or machines or tractors), and (4) the maximum number of hours each day that the tractor can be expected to be in the field. For example, assume the following: (1) 30 work days are available (Sundays are excluded) from April 15 to May 20, (2) 60% of these days are suitable for field work, (3) two tractors and two operators are available, and (4) each tractor can be in the field 11 hours per day. Then the B column coefficient would be $30 \times .60 = 18$ days, $18 \times 2 = 36$ tractor days, and $36 \times 11 = 396$ tractor hours available for field work.

R01 = a land restraint. The B column unit is acres.
R02 = a labor rstraint from April 15 to May 20. The B column unit is hours.
R03 = a restraint on tractor field time available during the period from April 15 to May 20. The unit is one hour of tractor time.
R04 = a transfer row transferring corn that has just been planted. The transfer unit is one acre.
R05 = a restraint on the amount of part-time help which can be hired.
P01 = preparing land in corn stalks and planting corn. The unit of activity is one acre.

P02 = a hog farrowing activity. The unit of activity is one sow and litter.
P03 = an activity which hires labor for field operations. The unit of activity is one hour.

	Restraint Description	B	P01	P02	P03
R01	Land	300	1		
R02	Labor	600	2.0	2	—.95
R03	Tractor field time Apr. 15–May 20	396	1.56		
R04	Planted corn transfer row	0	—1		
R05	Hired help	200			1

Points to Observe

1. The total amount of labor required during this period by P01 exceeds the amount of tractor field time required. The difference represents tasks such as procurement of supplies and preparation and maintenance of equipment which may be performed during periods not suitable for work in the field.
2. Activity P02 is added to remind the reader that under some conditions, activities not involving field work may compete for labor required by cropping activities.
3. This example assumes that two operators and two tractors are available. In some instances it may be possible to supplement B column operator time by hiring part-time help. P03 represents such an activity. The difference between the P03 coefficient in R02 and the P03 coefficient in R05 indicates that a small percentage of time is lost between arriving for work and reaching the field.
4. R03 represents tractor time because we determined that tractors were the most restricting type of equipment by our calculations in Table 6.1. These calculations must be repeated for each field time restraint that is formed.

Multiple Field Time Restraints

In this section we extend the concept of field time restraints to include three operations: (1) seedbed preparation, (2) planting, and (3) harvesting where timeliness is critical. The first step as before is to estimate the total number of hours of suitable field time available (Table 6.2). The information required to construct Table 6.2 is as follows:

1. The length of the period considered desirable from the standpoint of timeliness for performance of the operation in question. If more than one time period is included, the range of each must be specified. Including more than one planting period implies that one will be able to differentiate yield coefficients among periods included.
2. The number of hours each day that the operation can be carried out.
3. A determination on the part of the operator if Sunday field work is permissible.
4. An estimate of the number of days suitable for field work from the standpoint of weather and field conditions.

Once the suitable field time estimates have been made, the most limiting element must be identified as illustrated in Table 6.3, following the same pattern as shown in Table 6.1, only this time we are concerned with three phases of the production process instead of one. The results of this analysis are reported in Table 6.3.

The reader should note that whereas tractor power was the most limiting element in the field preparation phase, the capacity of the planter is limiting during the planting season and of the cornpicker during harvest.

The data we have developed are integrated into a partial model below.

MODEL 6.2: *Multiple Restraints on Suitable Field Time*

Explanation

This model differs from previous illustrations in that (1) it permits multiple restraints on suitable field time and (2) it does not assume

TABLE 6.2: Specification of Suitable Field Hours by Operation and Time Periods

Operation and Period	Hours of Labor Available	Suitable Field Days Excluding Sunday	Suitable Field Hours Excluding Sunday
Seedbed preparation Apr. 1–Apr. 20	374	10.2	112.2
Planting Period I Apr. 21–May 15	462	13.6	149.6
Planting Period II May 16–May 28	220	7	77
Harvesting Period I Oct. 25–Nov. 15	342	14.2	127.8
Harvesting Period II Nov. 16–Dec. 5	389	10.2	86.7

TABLE 6.3: Diagnosing the Limiting Element during Three Phases of Corn Production

Operation	Hours per Acre	No. of Machines	Field Time per Acre (hours)
Seedbed preparation:			
Fertilizer spreader	.12	1	.12
Disk (stalks)	.17	1	.17
Plow	.64	2	.32
Disk (plowed land)	.17	1	.17
Tractor	1.10	2	.55*
Planting:			
Disk	.17	1	.17
Planter	.29	1	.29
Tractor	.46	2	.23
Harvesting:			
Picking machine (mounted)	.45	1	.45
Hauling and unloading equipment	.32	1	.32
Drying equipment	.08	1	.08
Tractor	.77	2	.38

* Assumes overlapping of operations and hence full use of field time for each of the two tractors.

that either tractor time or operator time will be the most limiting element, as does Model 3.26.

P01 = a seedbed preparation activity. This includes plowing and disking. The unit of activity is one acre.
P02 = a corn planting activity for Period I. The unit of activity is one acre.
P03 = a corn planting activity for Period II. The unit of activity is one acre.
P04 = an activity which harvests corn in Harvest Period I that was planted in Planting Period I. The unit of activity is one acre.
P05 = an activity which harvests corn in Harvest Period II that was planted in Planting Period I. The unit of activity is one acre.
P06 = an activity which harvests corn in Harvest Period I that was planted in Planting Period II. The unit of activity is one acre.
P07 = an activity which harvests corn in Harvest Period II that was planted in Planting Period II. The unit of activity is one acre.
R01 = a restraint on total labor available for the period Apr. 1–20. The B column unit is hours.
R02 = a restraint on total labor available Apr. 21–May 15. The B column unit is hours.
R03 = a restraint on total labor available May 16–28. The B column unit is hours.
R04 = a restraint on total labor available Oct. 25–Nov. 15. The B column unit is hours.

MODEL 6.2

		B	Seed bed Preparation P01	Planting Period I P02	Planting Period II P03	Early Harvesting Period I Corn P04	Late Harvesting Period I Corn P05	Early Harvesting Period II Corn P06	Late Harvesting Period II Corn P07
C			—15.75	—16.00	—15.50	—2.35	—2.35	—2.35	—2.35
R01	Labor	374	1.10						
R02	Labor	462	.69	.46					
R03	Labor	220			.46				
R04	Labor	342				.90		.90	
R05	Labor	289					.90		.90
R06	Field time Apr. 1–Apr. 20	112.2	.55						
R07	Field time Apr. 21–May 15	149.6		.29					
R08	Field time May 16–May 28	77.0			.29				
R09	Field time Oct. 25–Nov. 15	127.8				.45		.45	
R10	Field time Nov. 16–Dec. 14	86.0					.45		.45
R11	Standing corn transfer I	0		—1		1	1		
R12	Standing corn transfer II	0			—1			1	1
R13	Corn grain transfer	0				—100	—98	—96	—93
R14	Land prepared for planting	0	—1	1	1				
R15	Land	400	1						

This model has not been developed to include land preparation from Dec. 15–Mar. 31 which is possible in many cases; also, land may be prepared during planting periods that have been omitted in this case.

R05 = a restraint on total labor available Nov. 16–Dec. 5. The B column unit is hours.
R06 = a restraint on suitable field time Apr. 1–20. The B column unit is hours.
R07 = a restraint on suitable field time Apr. 21–May 15. The B column unit is hours.
R08 = a restraint on suitable field time May 16–28. The B column unit is hours.
R09 = a restraint on suitable field time Oct. 25–Nov. 5. The B column unit is hours.
R10 = a restraint on suitable field time Nov. 16–Dec. 14. The B column unit is hours.
R11 = a standing corn transfer row for corn planted during Period I. The transfer unit is acres.
R12 = a standing corn transfer row for corn planted during Period II. The transfer unit is acres.
R13 = a corn grain transfer row. The transfer unit is bushels.
R14 = a transfer row for land prepared for planting. The transfer unit is acres.
R15 = a land restraint. The B column unit is acres.

Points to Observe

1. The most limiting element in seedbed preparation is tractor time. The model assumes that two tractors are available and that both can be used simultaneously in preparing land for planting.
2. The most limiting element during the planting season is the planter and during harvesting, the cornpicker. The coefficients at the intersection of the restraints on field time suitable for planting (rows R08 and R09) and the planting activities (P02 and P03) were developed in Table 6.3, as were those for the harvesting activities.
3. Restraints on labor are also included in the model. When livestock activities are included, they may compete with cropping activities during critical periods.
4. As constructed the model assumed that livestock activities would use labor first from those periods not suitable for field operations.
5. In this model only two planting and two harvesting activities have been included. These could be proliferated to any number desired, although those added beyond three or four probably will add little to the effectiveness of the model.
6. Land preparation and planting periods can overlap if appropriate coefficients can be defined. Land preparation can extend from April 20 through May 5 and planting from May 1 to May 10.
7. Where corn is planted during different periods, it must be transferred as standing corn in separate rows and harvested by a separate activity. The expected difference in yield arising from the different planting period can then be reflected in the coefficients at the intersection of the harvesting activities and the corn grain transfer row.

TIME-CRITICAL AND TIME-FLEXIBLE REQUIREMENTS

Farming operations may be classified into time-critical and time-flexible in respect to manpower, tractor power, and machine requirements. The concept is not new. Indeed, time-flexible operations have long been set aside as rainy-day jobs or slack-season work. Fence and building repair, manure disposal, and many procurement activities typically fall into the time-flexible category. On the other hand, seed-bed preparation, planting, weed control, and harvesting are time-critical operations because their postponement may result in substantial losses to the farm business.

There are few operations which do not contain some time-flexible components. Corn planting where timing is rightly considered to be critical contains some tasks that are time-flexible. Procurement of seed, proper adjustment of the planter, procurement of fertilizer, and even application of fertilizer if plow-down is used are examples of time-flexible tasks. A rigorous approach to farm planning requires that time-flexible and time-critical tasks be distinguished and separate labor restraints be specified for each type or model developed wherein time-flexible tasks can be postponed (or advanced) to another time period.

CHAPTER SEVEN

METHODS TO EVALUATE SENSITIVITY OF THE PLAN

A complete interpretation of a farm plan developed through linear programming requires investigation of the stability of the plan. The answers to questions like the following often provide useful insights into the planning situation.

1. How great is the advantage of activities which entered the plan over those which did not?
2. How would increasing or decreasing one or more resources affect the optimum mix of activities and the value of the program?
3. How would changes in price relationships affect the solution?

Useful information bearing on all three of these questions can be gleaned from a careful analysis of the shadow prices included in a conventional output report. The shadow prices on real activities not in the solution indicate by how much income would be penalized were they forced into the plan. Hence close competitors to those activities in the plan can be identified. However, the routine output report does not indicate the range over which the income penalty reported is relevant, because the point at which the penalty will increase is not reported in the output. Likewise, the shadow prices on disposal activities indicate the marginal contribution to income of the last unit of resource. They do not specify how many units of a resource, if any, could be added without lowering its marginal contribution.

The multiple B column option explained in Chapter 3 provides a device for analyzing the effect of resource changes on the plan. The optimum plan, the level of income, and the marginal value product of a resource can be estimated for any level of resource use or combination of restraints desired by adding appropriate multiple B columns to the model. The multiple C row option, also explained previously, provides insight into the sensitivity of the plan to changes in price relationships. The multiple B and multiple C devices can be implemented with little added effort, and both use computer time efficiently. One can define an infinite number of C rows, all representing reasonable combinations of price expectations. The same can be said for resource restraints. The planner must be discriminating in selecting the relationships he attempts to analyze lest he generate a greater volume of output than he can interpret.

TABLE 7.1: Range Analysis Output Report

SECTION 1 - ROWS AT LIMIT LEVEL

	Column 1		Column 2	Column 3	Column 4	Column 5	Column 6		Column 7
NUMBER	...ROW..	AT	...ACTIVITY...	SLACK ACTIVITY	..LOWER LIMIT. ..UPPER LIMIT.	LOWER ACTIVITY UPPER ACTIVITY	...UNIT COST.. ...UNIT COST..	..UPPER COST.. ..LOWER COST..	LIMITING PROCESS.
2	R01	UL	2400.00000	.	NONE 2400.00000	1387.20020 2737.57837	4.28525- 4.28525		P02 P08
4	R03	UL	150.00000	.	NONE 150.00000	129.77554 352.55998	56.57373- 56.57373		P08 P02
5	R04	UL	18.00000	.	NONE 18.00000	. 46.61015	35.90707- 35.90707		P04 P02
6	R05	UL	.	.	NONE .	2598.62573 INFINITY	1.20000- 1.20000		P08 NONE
7	R06	UL	.	.	NONE	38.71478- 28.13333	39.73093- 39.73093		P08 P02
9	R08	UL			NONE	717.04297- INFINITY	.67000- .67000		P11 NONE

SECTION 2 - COLUMNS AT LIMIT LEVEL

	Column 1			Column 2		Column 3	Column 4	Column 5	Column 6
NUMBER	.COLUMN.	AT	...ACTIVITY...	..INPUT COST..	..LOWER LIMIT. ..UPPER LIMIT.	LOWER ACTIVITY UPPER ACTIVITY	...UNIT COST.. ...UNIT COST..	..UPPER COST.. ..LOWER COST..	LIMITING PROCESS.
12	P03	LL	.	89.00000-	. NONE	16.65788- 4.50000	50.05669 50.05669-	INFINITY- 38.94331-	P02 P04
14	P05	LL	.	79.00000	. NONE	8.16774- .	171.62877 171.62877-	INFINITY- 250.638877	P02 R07
15	P06	LL	.	92.00000	. NONE	48.22856- .	84.00970 84.00970-	INFINITY- 176.00990	P02 R07
18	P09	LL	.	4.00000-	. NONE	. 43.72989	48.01617 48.01617-	INFINITY- 44.01617	R07 P08
19	P10	LL	.	1.25000-	. NONE	2957.52573- 14486.60937	.05000 .05000-	INFINITY- 1.20000-	P08 R02

SECTION 3 - ROWS AT INTERMEDIATE LEVEL

	Column 1		Column 2	Column 3	Column 4	Column 5	Column 6		Column 7
NUMBER	...ROW..	AT	...ACTIVITY...	SLACK ACTIVITY	..LOWER LIMIT. ..UPPER LIMIT.	LOWER ACTIVITY UPPER ACTIVITY	...UNIT COST.. ...UNIT COST..	..UPPER COST.. ..LOWER COST..	LIMITING PROCESS.
3	R02	BS	7205.35547	5794.64453	NONE 13000.00000	4728.00366 INFINITY	1.75191- .12500-		R01 P10
8	R07	BS	.	.	NONE .	43.72987- 47.74484	48.01617- 84.00970-		P09 P06

SECTION 4 - COLUMNS AT INTERMEDIATE LEVEL

	Column 1		Column 2	Column 3		Column 4	Column 5	Column 6	Column 7
NUMBER	.COLUMN.	AT	...ACTIVITY...	..INPUT COST..	..LOWER LIMIT. ..UPPER LIMIT.	LOWER ACTIVITY UPPER ACTIVITY	...UNIT COST.. ...UNIT COST..	..UPPER COST.. ..LOWER COST..	LIMITING PROCESS.
10	P01	BS	113.56834	30.00000-	. NONE	103.72662 149.99998	22.88776- 119.12999-	52.88776- 89.12999	P03 R01
11	P02	BS	9.10791	81.00000-	. NONE	3.69642 11.56834	476.52002- 91.55104-	557.52002- 10.55104	R01 P03
13	P04	BS	4.50000	58.00000-	. NONE	47.42761- 4.50000	50.05669- INFINITY-	108.05669- INFINITY-	P03 NONE
16	P07	BS	45.34532	446.00000	. NONE	16.20001 47.48201	148.91249- 302.48975-	297.08751 748.48975	R01 R04
17	P08	BS	2957.14746	1.20000	. NONE	798.56128 17085.23120 INFINITY	.35906- .05000- .67000-	.84094 1.25000	R04 P10
20	P11	BS	717.04297	.67000	NONE	854.82710	1.63484-	2.30484	

RANGE ANALYSIS

The range analysis extends the information provided in the conventional solution. It has the effect of making more useful the interpretation of the shadow prices by providing an estimate of the range over which a shadow price is relevant (see Table 7.1).

The range output can be obtained with ease with MPS/360 routine. A RANGE card is added to the control deck immediately following the SØLUTIØN card. No additional data cards or instructions are required of the user.

Interpretation of the range report is difficult, largely due to its unfamiliar terminology. The report contains four sections as follows:

1. Section 1, Rows at Limit Level, reports on the restraint rows where the slack (disposal) activity is at zero level. Thus resource restraints presented in this section are those fully used in the plan and therefore limiting.
2. Section 2, Columns at Limit Level, is concerned in the farm planning context with those activities which have been left out of the plan. They are at a lower limit of zero.
3. Section 3, Rows at Intermediate Level, provides an analysis of restraints with slack activities at nonzero level.
4. Section 4, Columns at Intermediate Level, analyzes real activities which are in the basis.

An example of the range analysis (Table 7.1), derived with the Winterset model, is shown in four sections and discussed section by section. The reader should study one section until he understands the interpretation of the example before moving to another. Interpretation of the range analysis on transfer rows and activities is tedious and unrewarding in terms of additional insight. Consequently, it is suggested that the student bypass interpretation of transfer rows and transfer activities during his first encounter with the range analysis.

Section 1—Rows at Limit Level

We focus our attention first on R01, a row (labor restraint) at the limit. In this case 2,400 hours of labor were entered in the B column of the original model (Winterset Model, Chapter 4), and all 2,400 hours are being used in the plan.

The columns of particular interest in this section are 5, 6, and 7. The first four appear in the standard output report and hence need no further explanation. Column 6, labeled unit cost, shows the shadow price or marginal value product for labor in R01. Column 5 (lower activity, upper activity) shows the range over which the shadow price of $4.28 is relevant. Each hour of reduction in labor from 2,400 to 1,387 hours would reduce the value of the program by $4.28. Each hour added beyond 2,400 to 2,737 would add $4.28 to the value of the program. Column 7 specifies the activities now in the basis that drop out at the lower and upper limits of the constant marginal value product range. In this case P02 (CCOM) drops out at the lower activity level and P08 (corn selling) at the upper activity level.

Turning now to R03, the B column of the original model specified an upper limit of 150 acres of Winterset silt loam soil. The standard output indicated a marginal value product for land of $56.57, and this estimate is repeated in the range report. In addition, the range report tells us that this estimate is appropriate and constant from 150 acres down to 129.77 acres and from 150 acres up to 352.56 acres.

Section 2—Columns at Limit Level

In this section we are dealing with real activities included in the model which did not enter the plan. They are at their lower limit of zero. Column 4 gives the income penalty for each activity. This column repeats the shadow price information for the nonbasic real activities that is supplied in the conventional output report. Looking first at P03 we observe that the income penalty (reported in column 4 of Section 2 as Unit Cost) is $50.06. Column 3 indicates that this income penalty is constant over a range of —16.65 to 4.50 units. Since a real activity carried on at a negative level is impossible, the real range of the constant income penalty is from 0 (the present level) to 4.50 units of activity. If we forced P03 into the plan beyond the 4.5 level, the income penalty would increase, but the range report supplies no information concerning the magnitude of the increase.

Column 5 indicates that P03 (which is a cropping activity where the output is transferred, resulting in a negative net price) would enter the plan at a level of 4.5 if the net price were raised from —$89.00 to —$38.94.

The range analysis in this example has been of no help in the case of either P05 or P06. The range specified in each case extends from zero to a negative activity level. We must treat the latter as meaningless on logical grounds. The zero upper limit is misleading. The income penalties as reported in both the conventional and the range output are $171.63 and $84.01. An increase in the C row coefficient from $79 to $250.63 (79 + 171.63) in the case of P05 and from $92 to $176.01 (92 + 84.01) would result in P05 and P06 entering the solution, the zero upper limit notwithstanding. Before P05 and P06 can enter the solution in response to higher C row coefficients, the basis must change to include hay production and harvesting, thus making R07 no longer limiting. The interpretation of the range output as it relates to P05 and P06 emphasizes the limitations of the range analysis where transfer rows are involved.

Section 3—Rows at Intermediate Level

This section is concerned with those restraints which are not limiting. Information on transfer rows may also be included, but this section of the report should be ignored. In the case at hand, only one restraint, R02 (capital), is not at the limit. Hence the slack activity entered the solution at a level of $5,794.64. This is the amount of capital which would go unused in the optimum plan. Changing the amount of capital used—either forcing the plan to make use of more or permitting less to be used—will reduce income.

Column 6 provides an estimate of the income penalty incurred in changing from the optimum amount of capital; column 5 indicates the range over which the income penalty extends. More specifically, on the up side any quantity of capital could be invested at a loss of $.125 per dollar. This arises because of an opportunity to engage in a losing corn buying–corn selling venture without limit. Of greater interest, the report also indicates in column 6 that reducing the amount of capital below the $7,205 used in the plan would involve an income penalty of $1.75 for each dollar withdrawn. This penalty would remain constant until capital use had diminished to $4,728, at which point the penalty presumably would become even greater.

Section 4—Columns at Intermediate Level

This section reports on the real activities which entered the plan. Because the plan is optimum, diverging from it will cause a decrease in the value of the program. We can diverge from the plan, in respect to any activity, by including the activity at a level either higher or lower than specified in the optimum plan. The range analysis of the real activities in the plan provides insight into the magnitude of the income penalties which attach to departing from the optimum on either the up or the down side.

P01 (continuous corn) is in the solution at a level of 113.56 acres. The first row of column 5 indicates that an income penalty of 22.89 arises for each acre the activity is decreased below the optimum. The same penalty applies until the acreage is decreased to 103.7 (from column 4, lower activity). Below this the penalty presumably increases. If P01 is pushed beyond the optimum, the penalty is $119.13 per acre. The same penalty applies up to 150 acres.

Column 6 provides an estimate of the sensitivity of P01 to changes in the C row coefficient. The reader will note that P01 has a C row coefficient of —$30. This coefficient can vary from —$52.88 upward to $89.13 before changes in the level of P01 in the solution would occur. At —$52.88 the level of P01 would drop to 103.7 acres; at $89.13 it would jump to 150, using all the land suitable for continuous row cropping.

PARAMETRIC ROUTINES

Parametric programming can be used to estimate the effect of either C row or B column changes on the optimum plan. It is particularly well adapted to planning situations where one wishes to know the effects of changes in resources or prices at many steps along the way. If one sought to program the effect of changes in the price of hogs from $16 to $22 and wanted a plan at each $.25 interval, the parametric routine is the appropriate device. Although it is possible to achieve the same result with the standard routine by making 24 separate runs or with a multiple C row model containing 24 C rows, the parametric approach is more efficient.

Changing One B Column Coefficient

A coefficient in the B column may be changed by a constant increment (or decrement) from the quantity defined in the original B column. The magnitude of the constant increment is specified by XPARDELT in the control deck. (Note the control cards in the example on page 126.) The procedure continues to increment the initial B column value until some maximum limit, specified by XPARMAX, has been reached. The model is optimized at each increment. In the Winterset example the original restraint on Winterset silt loam was 150 acres. Suppose now we wish to observe how the activity mix, shadow prices, and value of the program would change if one were to increase the amount of good land in 40-acre steps to 390 acres. XPARDELT is set at 40 and XPARMAX at 240 ($150 + 240 = 390$).

In addition to the definition of XPARDELT and XPARMAX in the control cards, the data deck must indicate the row to be changed. This requires a card containing the letters CHGCØL in columns 5–10 and the identification of the row being parameterized in columns 15–22. If one seeks to *increment* from the original B column, a coefficient of +1 is entered in the coefficients field (columns 25–36). A minus sign will result in negative changes of a magnitude defined by XPARDELT in the control deck.

MPS/360 Linear Programming
80 COLUMN DATA SHEET

PROGRAM			JOB NO.		BY	DATE	
1 Type 4	5 Name 12	15 Name 22	25 Coefficients 36	40 Name 47	50 Coefficients 61	70	80
NAME		ECØN430					
RØWS							
N	C						
L	R01						
L	R02						
----	---------NOTE:	RØWS L R03	- R08 HAVE BEEN OMITTED				
CØLUMNS							
	P01	C	-30.	R01	5.		
	P01	R02	20.	R03	1.		
	P01	R05	-90.				
----	---------NOTE:	P02 - P07 HAVE	BEEN OMITTED				
	P08	C	1.20	R05	1.		
	P09	C	-4.	R01	1.		
	P09	R02	3.	R06	1.		
	P09	R07	-1.				
	P10	C	-1.25	R02	.4		
	P11	C	.67	R08	1.		
RHS							
	B	R01	2400.	R02	13000.		
	B	R03	150.	R04	18.		
	CHGCØL	R03	1.				

FIG. 7.1. Data format for applying parametric feature to B column changes.

Changing More than One B Column Coefficient

To use the parametric device for changing more than one B column coefficient requires manipulation of the CHGCØL coefficient. In the previous example only the quantity of Winterset silt loam was changed. If one were analyzing the effect of added land on the optimum farm plan, it is unlikely that the additional land would all be Winterset silt loam. It would be more reasonable to assume that some portion of the land added would be the less desirable Shelby loam. If we assume 80% would be Winterset silt loam and 20% Shelby loam, the CHGCØL coefficient for R03 (Winterset silt loam) would be .8 and for R04 (Shelby loam) .2. Incrementing by XPARDELT (40) would cause a 32-acre ($.8 \times 40 = 32$) increase in R03 and an 8-acre ($.2 \times 40 = 8$) increase in R04.

If the ratio of R03 to R04 were to remain fixed at each step in a ratio equal to that defined by the original B column values, then the

PARAMETRIC ROUTINE (B Column)

Job Control Language and Control Program Cards

```
JOB CONTROL LANGUAGE CARDS


             PROGRAM
             INITIALZ
             MOVE(XDATA,'ECON430')
             MOVE(XPBNAME,'PBFILE')
             MVADR(XMAJERR,UNB)
             MVADR(XDONFS,NOF)
             CONVERT
             SETUP('MAX')
             MOVE(XRHS,'B')
             MOVE(XOBJ,'C')
             PRIMAL
             SAVE
             SOLUTION
             XPARAM=0.0
             MOVE(XCHCOL,'CHGCOL')
             XPARDELT=40.0
             XPARMAX=240.0
             PARARHS
             SOLUTION
             EXIT
NOF          TRACE
UNB          EXIT
             PEND
/*


JOB CONTROL LANGUAGE CARDS


NAME                ECON430


ENDATA
/*
```

CHGCØL coefficient for R03 should be .89275 and for R04 .10725. This follows because 150/168 or 89.275% of the total land in the original tract is Winterset silt loam and 18/168 or 10.725% is Shelby loam. The CHGCØL coefficients thus defined will result in a (40)(.89275) = 35.71 increment in R03 and a (40)(.10725) = 4.29 increment in R04 for each XPARDELT increment.

A similar procedure may be used to increment resources other than land. Assume, as in the previous example, XPARMAX = 240 and XPARDELT = 40; and we wish to increment capital by $2,000 for each 40-acre increase in land, moving in steps of $2,000 from $13,000 to $25,000. Then the CHGCØL in coefficient R02 will be 50 since (50)(40) = 2,000 and (240)(50) = 12,000. Thus capital can be incremented with land in any magnitude desired by manipulation of the CHGCØL coefficients.

Changes in the Objective Function

Implementation of the procedure for parameterizing the objective function is much the same as for the B column. The control deck is similar. The XPARDELT card specifies the constant increment (or decrement) by which one seeks to change one or more coefficients in the C row. The XPARMAX defines a maximum limit to

MPS/360 Linear Programming
80 COLUMN DATA SHEET

PROGRAM | JOB NO. | BY | DATE

1 Type 4	5 Name 12	15 Name 22	25 Coefficients 36	40 Name 47	50 Coefficients 61	70 80
NAME		ECØN430				
RØWS						
N	C					
N	CHGRØW					
L	R01					
L	R02					
---	--------------	--NOTE: RØWS CARDS L R03 - L R08 HAVE BEEN OMITTED				
CØLUMNS						
	P01	C	-30.	R01	5.	
	P01	R02	20.	R03	1.	
	P01	R05	-90.			
---	--------------	--NOTE: P02 - P07 HAVE BEEN OMITTED				
	P08	C	.95	R05	1.	
	P08	CHGRØW	1.			
	P09	C	-4.	R01	1.	
	P09	R02	3.	R06	1.	
	P09	R07	-1.			
	P10	C	-1.00	R02	.4	
	P10	R05	-1.			
	P10	CHGRØW	-1.			
	P11	C	.67	R08	1.	
RHS						
	B	R01	2400.	R02	13000.	
	B	R03	150.	R04	18.	

FIG. 7.2. Data format for applying parametric feature to objective function changes.

the amount of change. Thus, if one seeks to analyze the influence of increasing the selling price of corn from a C row price of $.95 in $.10 increments to a maximum of $1.45, the XPARDELT value would be .10 and the XPARMAX value .50. The same XPARDELT and XPARMAX can be used for a simultaneous change in the buying price of corn from $1.00 to $1.50.

PARAMETRIC ROUTINE (C Row)

Job Control Language and Control Program Cards

```
JOB CONTROL LANGUAGE CARDS


          PROGRAM
          INITIALZ
          MOVE(XDATA,'ECON430')
          MOVE(XPBNAME,'PBFILE')
          MVADR(XMAJERR,UNB)
          MVADR(XDONFS,NOF)
          CONVERT
          SETUP('MAX')
          MOVE(XRHS,'B')
          MOVE(XOBJ,'C')
          PRIMAL
          SAVE
          SOLUTION
          XPARAM=0.0
          MOVE(XCHROW,'CHGROW')
          XPARDELT=.10
          XPARMAX=.50
          PARAOBJ
          SOLUTION
          EXIT
NOF       TRACE
UNB       EXIT
          PEND
/*


JOB CONTROL LANGUAGE CARDS


NAME          ECON430


ENDATA
/*
```

The data deck must contain a card in the rows section identifying the CHGRØW as an N type row. The column (or columns) whose C row coefficients are to be changed must be identified in the columns section. In the example, C row coefficients for P08 and P10 are to be changed; consequently both appear in the column identification section. P08 is a corn selling activity with a positive sign, and its price is to increase in \$.10 increments (XPARDELT = .10); consequently its CHGRØW coefficient is 1. On the other hand, P10 is a corn buying activity with a negative coefficient in the C row. To cause the buying price to increase in \$.10 increments (become more negative) one must enter a coefficient of −1 (instead of a + 1) for the P10 coefficient in row CHGRØW.

The prices on other activities may be changed along with corn prices and in any ratio desired.

Assume, as in the previous example, XPARMAX = .50 and XPARDELT = .10; and we wish to increment the C row value on the hog raising and selling activity (P07) by \$30 for each \$.10 increase in the price of corn, in steps of \$30 from \$446 to \$596. The the P07 coefficient in CHGRØW should equal 300, since (.50)(300) = 150, the *total* increase in the C row value desired in P07, and (.10)(300) = 30, the step increase sought in the C row coefficient in P07.

CHAPTER EIGHT

OTHER MODIFICATION PROCEDURES

We have already discussed a number of procedures for modifying the model after the initial solution to extend the range and usefulness of the information forthcoming. Among these are multiple B columns, multiple C rows, and parametric routines. To this inventory of analytical tools we now add control statement REVISE with the types of revision MØDIFY, DELETE, and BEFØRE and AFTER. These are presented because they (1) are more general and offer greater flexibility to the user since they permit making within-run changes of any coefficient in the model and (2) make possible conditional revisions which in turn clear the way for several new and highly useful programming alternatives. The two classes of revisions, unconditional and conditional, are described first. Then the types of revisions, all of which may be applied to either class, are presented.

UNCONDITIONAL REVISE

The unconditional revise procedure permits any number of coefficients within the model to be changed prior to reoptimization. The procedure is unconditional because the nature and magnitude of the changes desired are fully specified prior to the initial optimization and are in no way a function of its outcome. The unconditional revise could be used to change the yield coefficients of corn and soybeans after one optimization in anticipation of another. Any number of reoptimizations can be called for. The yield of corn could be changed step by step from 90 to 96 to 100 to 102 and so forth to include any number of revisions.

The control program for applying unconditional revise procedures is given on page 131. The types of revisions available and the data format for each type are described under the section entitled Types of Revisions.

The statements in the program listing HIP MØVE(XØLDNAME, 'PBFILE') through SØLUTIØN revise the prior model according to the data supplied by the user and then proceed to optimize the revised model. VPAR = VPAR —1 through GØTØ(HIP) and VPAR

DC(2) statements serve as a counter which transfer control (when multiple revisions are desired) to the second and successive revise or to the end. Activities can be added and/or deleted and coefficients within existing activities can be modified. Furthermore, any number of coefficients within the model can be changed prior to reoptimization. Thus one could change the yield coefficients of corn and soybeans (or

UNCONDITIONAL REVISE

Job Control Language and Control Program Cards

```
JOB CONTROL LANGUAGE CARDS


            PROGRAM
            INITIALZ
            MOVE(XDATA,'ECON430')
            MOVE(XPBNAME,'PBFILE')
            MVADR(XMAJERR,UNB)
            MVADR(XDONFS,NOF)
            CONVERT
            SETUP('MAX')
            MOVE(XRHS,'B')
            MOVE(XOBJ,'C')
            PRIMAL
            SAVE
            SOLUTION
HIP         MOVE(XOLDNAME,'PBFILE')
            REVISE
            SETUP('MAX')
            RESTORE
            PRIMAL
            SAVE
            SOLUTION
            VPAR=VPAR-1
            IF(VPAR.LE.1,HOP)
            GOTO(HIP)
HOP         EXIT
VPAR        DC(2)
NOF         TRACE
UNB         EXIT
            PEND
/*

JOB CONTROL LANGUAGE CARDS


NAME            ECON430


ENDATA
/*
```

any number of activities) for the second optimization. For example, where the yield of corn is changed step by step from (say) 90 to 96 to 100 to 102, four optimizations would be forthcoming—one from the original model and one from each of the three revisions.

CONDITIONAL REVISE

Unlike the unconditional revise where the type and magnitude of revisions are predetermined and revisions are performed regardless of the results of the initial optimization, the conditional revise routine examines one or more parameters of the first solution to determine if a revision is appropriate. Thus the nature of the revision is conditioned by the results of the previous optimization.

Again we return to the Winterset model for an example with which to demonstrate the conditional revise routine. In this model certain production coefficients were specified. These were based on the premise that the activities would be conducted on a scale where coefficients would not be affected unduly by changes in scale. Thus the beef cow-calf, the yearling steer feeding, and the swine activities should enter the program either not at all or at a level where the average cost curve has flattened out. We specify on the basis of what information we can gather that in the beef cow-calf activity this point occurs at 20 head. Therefore, we arrange for the activity to be dropped if, after optimization, the activity level is greater than zero but less than 20 head. The control program listed for the conditional revise is shown on page 134. The A1 DC(0.0), A2 DC(0.0), and A3 DC(20.0) cards define the constants A1, A2, and A3 to 0.0, 0.0, and 20.0 respectively, against which the level of P05 can be tested. The SELECT('CØL','P05',A1,' ') statement sets the constant A1 equal to the level at which P05 entered the first solution. With A1 equal to the activity level in the initial solution, we arrange to compare it with the levels that we set as boundaries. The IF(A1.GT.A3,HØP) statement tests the activity level A1 against the upper boundary A3, sending the program to the end if the number of cows is greater than 20. The next IF statement tests the lower boundary transferring control to the end only if A1 equals A2 or zero. If the activity level is greater than zero but less than 20.0, the two conditions will not be met, resulting in a revised solution where the beef cow activity will be deleted from the model. (See the data set in Fig. 8.1.)

The conditional revise program could also be used to add a labor hiring activity, should labor become a limiting resource in the initial solution. In this case the A1 DC(0.0) and A2 DC(2400.0)

MPS/360 Linear Programming
80 COLUMN DATA SHEET

PROGRAM				JOB NO.	BY	DATE

1 Type 4	5 Name 12	15 Name 22	25 Coefficients 36	40 Name 47	50 Coefficients 61	70	80
NAME		ECØN430					
RØWS							
N	C						
L	R01						
	Remaining rows have been omitted to conserve space.						
CØLUMNS							
	P01	C	-30.	R01	5.		
	P01	R02	20.	R03	1.		
	P01	R05	-90.				
	P02-P09 have been omitted to conserve space.						
	P10	C	-1.25	R02	.4		
	P10	R05	-1.				
RHS							
	B	R01	2400.	R02	13000.		
	B	R03	150.	R04	18.		
ENDATA							
NAME		ECØN430					
CØLUMNS							
DELETE							
	P05						
ENDATA							
	This revise data set will delete activity P05.						

FIG. 8.1. Format of a REVISE data set for deleting an activity.

are specified with A1 equal to any decimal number and A2 equal to the B column labor coefficient (R01). A SELECT ('RØW', 'R01', A1,' ') statement would set A1 equal to the amount of labor used in the initial optimization. If all the labor in the initial B column (in this case 2,400 hours) were not used, an IF (A1.NE.A2, NOG) would transfer the program to the end because a revision and an additional solution would be unnecessary.

The conditional revise procedure illustrated above is best adapted to models where only one parameter is tested in making the decision of whether to revise or to end the data processing. Source programs, using MPS Marvel, Read Comm, or Report Generator languages, can be written to test multiple parameters. These provide for repeated control transfers to develop and revise the data set internally during processing.

With the MPS routines it is feasible to perform a series of tests on the prior solution and arrange the type and number of revisions desired. To illustrate the degree of flexibility Marvel, Read Comm,

CONDITIONAL REVISE

Job Control Language and Control Program Cards

```
JOB CONTROL LANGUAGE CARDS


          PROGRAM
          INITIALZ
          MOVE(XDATA,'ECON430')
          MOVE(XPBNAME,'PBFILE')
          MVADR(XMAJERR,UNB)
          MVADR(XDONFS,NOF)
          CONVERT
          SETUP('MAX')
          MOVE(XRHS,'B')
          MOVE(XOBJ,'C')
          PRIMAL
          SAVE
          SOLUTION
          SELECT('COL','P05',A1,' ')
          IF(A1.GT.A3,HOP)
          IF(A1.EQ.A2,HOP)
          MOVE(XOLDNAME,'PBFILE')
          REVISE
          SETUP('MAX')
          RESTORE
          PRIMAL
          SOLUTION
HOP       EXIT
NOF       TRACE
UNB       EXIT
A1        DC(0.0)
A2        DC(0.0)
A3        DC(20.0)
          PEND
/*


JOB CONTROL LANGUAGE CARDS


NAME          ECON430


ENDATA
/*
```

or Report Generator provide, assume we wish to delete a cow-calf activity if its level in the solution is less than 20; but if the activity level is greater than 20, we want to modify the input coefficients to correspond to the level at which the activity entered the first solution. Such a modification can be accomplished without interrupting the computer run. Furthermore, we can arrange similar modifications for any number of activities in the model.

The entire range of the data set may be made conditional upon the outcome of the first or a previous optimization. The components of the model that are subject to revision and the nature of the criteria for revision must be prespecified. The criteria for deleting or modifying coefficients may be different for each coefficient.

The relational operators available in the system are:

.LT.	Less Than
.GT.	Greater Than
.EQ.	Equal To
.NE.	Not Equal To
.GE.	Greater Than or Equal To
.LE.	Less Than or Equal To

The types of revisions that can be accomplished under both the unconditional revise and the conditional revise programs are discussed below with examples of the appropriate data form.

TYPES OF REVISIONS

Modify

MØDIFY may be used to change the value of any coefficient within the CØLUMNS, RHS, or BØUNDS sections. Thus the planner may change any numerical information and cause reoptimization once the first solution has been obtained. The change may be predetermined and hence independent of the results of the first optimization, or it may be a function of the first solution and hence conditional. In addition, the type of restraint (N, G, L, E) may be changed in the RØWS section, or the type of bound (LO, UP, FX) may be changed.

Should the need arise to change to zero a nonzero coefficient from the original model, the zero must appear (cannot be an implied zero) in the revise data or the change will not be processed. The reader will recall that this entry of the zero in the MØDIFY routine departs from the usual practice in preparing data for processing where zeros are implied.

Delete

The DELETE procedure may be used to drop any activity or restraint from the initial model prior to reoptimization. One important application of this procedure is the tempering of the most dam-

MPS/360 Linear Programming
80 COLUMN DATA SHEET

PROGRAM				JOB NO.	BY	DATE
1 Type 4	**5 Name 12**	**15 Name 22**	**25 Coefficients 36**	**40 Name 47**	**50 Coefficients 61**	**70 / 80**
NAME		ECØN430				
CØLUMNS						
AFTER						
	P12	C	-35.0	R01	-1.	
ENDATA						
This revise data set will add activity P12 at the end of the model.						
NAME		ECØN430				
CØLUMNS						
BEFØRE		P11				
	P11A	C	-.75	R08	-1.	
ENDATA						
BEFØRE P11 will add the activity P11A before the activity P11 in the model.						
NAME		ECØN430				
CØLUMNS						
MØDIFY						
	P07	R01	30.	R05	240.	
	P07	R06	0.			
ENDATA						
The MØDIFY will change those elements P07 - R01, R05 and R06.						

FIG. 8.2. Format of REVISE data set for adding and modifying activities.

aging errors growing out of the linearity assumption. Coefficients in the original model can be estimated on the assumption that the activity will enter at a predetermined level considered realistic. If this condition is not met, the activity can be eliminated and a new solution obtained with little effort on the part of the user.

The DELETE routine is also useful when the planner wishes to ascertain how much a particular activity (or enterprise) adds to the value of the program. This is accomplished by optimizing first with the enterprise included among the possible activities and then reoptimizing with the activity deleted.

Before and After

The BEFØRE and AFTER routines are used to add activities or restraints to the model after an initial optimization has been performed. The only difference between BEFØRE and AFTER routines is the position assigned the activity when it is added to the model. An example will help to demonstrate the uses of the statement forms.

BEFØRE	P09			
P09A	C	—20	R07	—1
BEFØRE				
P09A	C	—20	R07	—1

The first statement form says add activity P09A before activity P09 in the model. The second statement says add P09A before the first activity in the model.

The AFTER statements are similar in form but place the activity at the end of the model unless another position is specified, e.g., AFTER P01.

Where rows are to be added, the BEFØRE or AFTER routine is used to establish the row in the RØWS section of the data set. Then the MØDIFY routine changes the implied zero elements to those actually desired in the new rows of the CØLUMNS section. Here the user must be careful to define the row to be changed by a BEFØRE or an AFTER statement prior to defining the coefficients with MØDIFY. Any combination of BEFØRE, AFTER, DELETE, and MØDIFY statements may be used in a single REVISE.

Preparation of the Data Deck

The order of the DATA deck is as follows:

1. *The initial data* set which is defined in the usual manner with NAME, RØWS, CØLUMNS, RHS, BØUNDS (if desired), and ENDATA.
2. *The revise data* set including
   a. a name card which must be an exact duplicate of the name on the initial data set,
   b. the name of the section to be altered which will be RØWS, CØLUMNS, RHS, or BØUNDS,
   c. the type of REVISE instruction desired (MØDIFY, DELETE, BEFØRE, or AFTER),
   d. the data relevant to the revision,
   e. the ENDATA card.

CHAPTER NINE

FORMING CAPITAL RESTRAINTS AND COEFFICIENTS

Forming meaningful capital restraints and coefficients is perhaps the most vexing problem one encounters in constructing a programming model. For this reason you should be thoroughly familiar with other facets of model building before attempting to incorporate capital restraints and coefficients into the model. Unfortunately there is no wholly satisfactory method of treating capital. Several alternatives are discussed below. The one you select depends on your experience in model building, the time and effort you are prepared to give to the task, and the precision your planning situation demands.

MODELS WITH NO CAPITAL RESTRAINTS

In some situations one may be confident that capital will not limit the plan, because other restraints—either real or subjective—are more limiting and hence determining. Under these circumstances the same mix of activities would result from a model with no capital restraints as from one with elaborate provisions for treating capital. Use of the no-capital restraint method implies the operator is willing and able to continue investing capital in the business as long as doing so will add to income.

The output from a model with no capital coefficients will not provide an estimate of the amount of capital required to carry on the program. Should the planner using this method desire estimates of capital needs, the latter must be prepared by budgeting.

MODELS FOR CAPITAL ACCOUNTING

In the accounting model we attempt to estimate the largest amount of capital the business will demand at the peak point during the production period. Presumably, if the operator can meet this requirement he will have sufficient capital during other periods of the production cycle. To permit forming only one capital coefficient per activity we limit our attention to the period of the year at which we predict capital will peak. Because we do this before we have optimized or obtained a solution, there is an element of guesswork involved. We assume that there is sufficient similarity among likely optimum farming plans to permit a realistic prediction of the period of peak capital needs. In Corn Belt agriculture, July would appear to be the critical month from the standpoint of capital.

The solution to a capital accounting model indicates how much capital will be required but provides no information on the optimum use of scarce capital. The latter answer requires that we include information concerning capital availability. Problems encountered in extending the model to take into account restraints on capital use are discussed later.

To illustrate the development of a model providing for peak period capital accounting, we return to the Winterset example. At this point you should disregard the capital coefficients found in the Winterset model in Chapter 4; in this chapter we develop a new set of coefficients based on July capital requirements.

MODEL 9.1: *Capital Accounting, Winterset Example*

Explanation

1. The restraints in the model are:

R01 = labor restraint. The B column entry is hours.
R02 = July peak period capital transfer row.
R03 = land restraint, Winterset silt loam. The B column entry is acres.
R04 = land restraint, Shelby loam. The B column entry is acres.
R05 = corn grain transfer row. The transfer unit is one bushel.
R06 = standing meadow transfer. The transfer unit is tons of hay equivalent.
R07 = hay transfer row. The tansfer unit is tons.
R08 = oats grain transfer row. The transfer unit is bushels.

2. The activities in the model are defined as follows:

P01 = continuous corn on Winterset silt loam growing and harvesting. The activity unit is one acre.
P02 = CCOM on Winterset silt loam growing and harvesting. The activity unit is four acres.

P03 = CCOM on Shelby loam growing and harvesting. The activity unit is four acres.
P04 = COMM on Shelby loam growing and harvesting. The activity unit is four acres.
P05 = maintaining beef cow-calf enterprise. The activity unit is one beef cow.
P06 = feeding yearling steers. The activity unit is one steer.
P07 = farrowing, raising, and selling hogs. The activity unit is one sow.
P08 = corn selling. The activity unit is one bushel.
P09 = hay harvesting. The activity unit is one ton.
P10 = corn buying. The activity unit is one bushel.
P11 = oats selling. The activity unit is one bushel.
P12 = capital accounting.

3. The capital coefficients in this version of the Winterset example assume initially that facilities and equipment are available to accommodate any of the activities at the level at which they are likely to enter the plan. This may not be a realistic assumption, but it is the least complicated situation from which to begin the study of capital coefficients.
4. The restraints are the same as in the Winterset model with these two exceptions: (a) R02 is no longer a capital restraint but serves only as a capital transfer row; (b) the **B** column values for R01, R03, and R04 have been increased.
5. The activities in the model are the same with the exception of the addition of a capital accounting activity, P12.
6. The coefficients are the same except for those in the capital transfer row.
7. The timing of P06 and P07, not specified previously, becomes relevant in forming capital coefficients. The yearling feeder cattle in P06 are purchased in late November and sold in early August. In P08 the first litter farrows in February and is marketed in August; the second litter farrows in early September and is marketed in March.
8. The capital coefficients for each activity refer only to the amount of capital required by a unit of activity on July 15.
9. Only the capital that is a function of the level of the activity and can be directly allocated to it is specified in the capital coefficient.
10. Capital coefficients for cropping activities are formed on the basis of the amount of capital committed to the crop from the initial phase of the activity, including soil preparation, until July 15. In the continuous corn activity, P01 includes growing costs of $11.50 plus $15.00 per acre for fertilizer for

a total of $26.50. This estimate does not include interest or expenditures for labor.

11. Estimated capital requirements for P02, P03, and P04 are as follows:

	P02	P03	P04
Corn growing costs	$49.00	$48.00	$23.00
Oats growing costs	4.10	4.10	4.10
1st year meadow growing cost	3.30	3.30	3.30
2nd year meadow growing cost	..	..	0.00
Total	$56.40	$55.40	$30.40

12. The hay harvesting activity has a capital requirement derived from variable costs incurred in hay harvesting prior to July 15. The hay harvesting capital coefficient does not include the investment in machinery. Comparisons of alternative systems of hay handling or hay versus silage require separate optimizations, each with a different machinery capital coefficient. Models incorporating these characteristics are discussed later.

    The coefficients for haymaking assume expenses have been incurred for two out of three cuttings by July and that 75% of the expected yield has been handled. The capital coefficient is $3.74 per ton.

13. The formation of capital coefficients for livestock follows the same pattern as for crops; i.e., coefficients are based on the investment which has been made in the activity since the beginning of the production cycle. Although the Winterset example does not include a dairy activity, the problems peculiar to the formation of capital coefficients for dairy enterprises are included for readers who wish to extend the model to include dairy production. In the case of dairying the capital coefficient encompasses only the value of the cow, the young stock needed to provide her replacement, and a small inventory of feed. The dairy enterprise is unique in that it returns variable capital inputs almost immediately. Where July is selected as the peak period month, the feed inventory typically does not include hay or silage. The amount of feed inventory built into the capital coefficient is a function of the feed inventory you would expect the operator to have on hand at the peak capital point. A one-day supply probably is insufficient, but a two-week inventory would seem adequate for most circumstances.

MODEL 9.1

Row Type		B	P01	P02	P03	P04	P05	P06	P07	P08	P09	P10	P11	P12
N	C		—30	—81	—89	—58	79	92	446	1.20	—4	—1.25	.67	0
L	R01	4,800	5	16	16	12	20	15	36	0	1	0		
L	R02	0	26.50	56.40	55.40	30.40	220	228	207	0	3.74	0		—1
L	R03	300	1	4										
L	R04	36	0		4	4								
L	R05	0	—90	—176	—124	—66	5	60	210	1	0	—1		
L	R06	0	0	—3.2	—1.8	—3.6	4	1	1	0	1	0		
L	R07	0	0	0	0	0	2	1	0	0	—1	0		
L	R08			—56	—46	—46							1	

14. P05 (beef calf raising and selling) raises the problem of the point in time at which the buildup in capital requirements should begin. The cow represents the major component in the capital coefficient. By July, however, additional inputs of capital principally in the form of feed have been invested. Pasture is ignored since the capital coefficients for pasture production would include a component for any capital required. The capital coefficient shown in the model includes the feed required for wintering. It is assumed that no other capital commitments in the form of variable costs are made in the activity prior to July 1.
15. The capital coefficient for P07 consists of (a) the value of the sow and (b) the supplement and grain required to carry the sow and litter from breeding to July 15, plus a small outlay for other variable costs such as veterinary fees. Since the timing of this activity is such that the first litter will be nearing market weight as of July 15, we have used 45% of the total feed grain and variable costs for the activity unit in estimating the capital coefficient.
16. The capital coefficient for feeder cattle is formed in a similar manner. It includes the investment in the original animal plus feed and variable costs accumulated to July 15.
17. P08 (corn selling) has no capital requirement.
18. No capital coefficient is entered in P10, because capital invested in corn purchased and still tied up in July would appear as part of the feed or livestock inventory incorporated in the livestock coefficients.

Points to Observe

1. The level at which P12 enters the solution will provide an estimate of the amount of capital the plan would require in July. However, only those items which entered into the formation of the capital coefficients will be included in this estimate. Capital for items such as machinery, buildings, and equipment—which were assumed to be on hand and adequate—constitute an additional requirement.
2. Constructing and optimizing this model provides no information as to the availability of the capital needed to implement the plan. The planner can estimate capital available from all sources—his own, borrowed, and accounts payable—and then decide if the plan is realistic. If not, he can proceed to impose limits as suggested in the following pages.

3. This model assumed adequate crop machinery and livestock equipment were available at the start of the plan. Another starting point in forming capital coefficients is to assume that crop machinery but no livestock facilities are available. In this case one would include a component in the capital coefficient reflecting the investment in livestock equipment and facilities. Problems arise with buildings and some types of facilities in that the investment required per unit of activity declines as the level of activity increases. The only manner in which this problem can be treated is through a process of trial and error involving repeated optimizations. If activities enter at an unusually low level and hence have an unrealistically low investment for that size of enterprise, the coefficients can be adjusted upward to conform to realistic estimates. Following this adjustment the model is reoptimized. Conceivably such an approach could involve an interminable number of optimizations, but in practice the enterprise usually will enter (a) at the level where the coefficients for buildings and equipment are reasonable, (b) at a level so low that the activity can be deleted in a second optimization, or (c) not at all. Routines are also available wherein coefficients can be adjusted and the model reoptimized within a single computer run (see Chapter 8).
4. A third situation in respect to building and equipment capital frequently is encountered. The farm being programmed has housing, equipment, and facilities to accommodate an activity to a known and potentially limiting level. Beyond this level additional investment would be required. In this situation two activities are necessary, one for producing with the existing facilities and a second for any expansion. The first activity would be restrained to the level of existing capacity, and its capital coefficients would reflect the fact that facilities are available. The second activity would be the same as the first except that its capital coefficients would reflect the need for additional investment in facilities.

MODELS INCLUDING CAPITAL RESTRAINTS

The model illustrated in the previous section treated capital to the extent of estimating only the amount required during July, the anticipated peak month. However, if capital is likely to be limiting and we wish to optimize within a capital restraint, we must include within the model a reasonable estimate of the amount of capital available, i.e., estimate a **B** column entry.

To help organize the process of estimating capital restraints and coefficients, we can begin by preparing a simple ledger in the manner suggested in Table 9.1. Once this is done, we should scrutinize both sides of the ledger to ascertain which items can be eliminated from the model. If the quantities in the solution will ultimately weigh equally on both sides of the equation, nothing is gained by including them. For example, fertilizer dealers sometimes supply fertilizer with the expectation that they will receive no payment until after harvest. In other words, the dealer serves as a source of capital to help the operator over the capital peak in July. The interest cost is included implicitly in the price. Fertilizer capital need not be added into the B side of the equation as a source of capital nor appear in any of the capital coefficients for activities using fertilizer under this arrangement.

Additionally, for the sake of simplicity we may be willing to assume that we have enough machinery, equipment, or livestock housing to accommodate any plan likely to evolve from the programming solution. We begin by circling items where availability and requirement will either be equal or assumed to be equal on both sides of the

TABLE 9.1: Winterset Example: Capital Availability and Requirement Ledger

B Column	Side of Equation	Activity Coefficient Side of Equation
$ 4,000	Corn on hand	Starter fertilizer
1,400	Other grain on hand	Nitrogen side-dressing
4,000	Value of hogs on hand	Supplement inventory
0	Value of beef cows in the herd	Fences
0	Value of dairy cows in the herd	Buildings
0	Value of young dairy stock in the herd	Livestock
	Value of tractors	Livestock equipment
	Value of crop machinery	Tractor fuel
200	Supplies, fuel, and miscellaneous on inventory	Seed corn
100	Supplement inventory on hand	Feeder cattle
	Livestock equipment on hand	Brood sows
	Buildings, fences, water system	Beef cows
	Dealer credit available for starter fertilizer	Veterinary fees
		Repairs
	Dealer credit available for nitrogen side-dressing	Crop machinery
300	Accounts receivable minus accounts payable*	Tractors
$10,000	Total B column entry	

* Includes only those accounts which will be paid or received prior to July 1.

equation. The circled items enter neither the B column nor the capital coefficients of the activities, although any costs associated with these items are reflected in the C row. Then we add the B column side of the ledger. The sum represents the appropriate entry in the B column of the capital row.

MODEL 9.2: *July Capital Restraint and Borrowing Activities*

Explanation

1. The activities in this model are the same as in Model 9.1, except that P12 (capital accounting activity) has been deleted and P13 and P14 (capital borrowing activities) have been added.
2. P13 provides for borrowing capital to supplement the July capital restraint. The interest rate is 8 percent. This assumes that to loosen the July restraint, capital must be borrowed for a year. Alternative assumptions can be made concerning the period of borrowing. The interest rate should be adjusted to reflect shorter borrowing periods. P13 does not include accounts payable such as fertilizer to be paid after harvest. Fertilizer applied on this basis does not appear as a component of the capital coefficient in the relevant activities.
3. P14 is a borrowing activity for feeder cattle. It is treated as a separate activity because lenders typically are willing to extend credit for the purchase of feeder cattle on a less restrictive basis than for other purposes. Where feed is available, the line of credit for the purchase of feeder animals usually is not limiting. Thus the purchase of feeder cattle typically does not compete with other activities for capital.
4. R02 is a July capital restraint. The B column value is estimated as shown in Table 9.1. Machinery, equipment, and buildings have been excluded (indicated by the fact that they are circled in Table 9.1) both from the B column value and in forming coefficients.
5. R09 is a transfer row for feeder cattle capital. Note that its value is zero in the B column and —1 in column P14.
6. R10 restrains the production credit that can be borrowed. The restraint does not apply to credit for the purchase of feeder cattle. The example assumes that none of the capital (R02) in the B column entry is borrowed. Thus R10 repre-

MODEL 9.2

Row Type		B	P01	P02	P03	P04	P05	P06	P07	P08	P09	P10	P11	P13	P14
N	C		—30	—81	—89	—58	79	92	446	1.20	—4	—1.25	.67	—.08	—.04
L	R01	4,800	5	16	16	12	20	15	36	0	1	0			
L	R02	10,000	26.50	56.40	55.40	30.40	220	78	207	0	3.74	0		—1	
L	R03	300	1	4											
L	R04	36	0		4	4									
L	R05	0	—90	—176	—124	—66	5	60	210	1	0	—1			
L	R06	0	0	—3.2	—1.8	—3.6	4	1	1	0	1	0			
L	R07	0	0	0	0		2	1	0	0	—1	0			
L	R08			—56	—46	—46							1.		
L	R09	0		0				150							—1
L	R10	5,000												1	

sents an estimate of both the total and the additional credit available aside from dealer accounts and feeder cattle loans.

Points to Observe

1. There are two capital coefficients for P06 (feeder cattle), one in R02 and a second in R09. The component of the capital coefficient resulting from the investment in feed, feed inventory, and other variable costs is shown in R02. The original cost of the animal, that which would be financed through feeder cattle credit, appears in R09.
2. The C row entry for P13 and P14 should reflect the interest charge for that part of the year for which the money will be borrowed. In the case of P14 the credit likely will be extended for the length of the feeding period. It is more difficult to estimate the length of loan (and hence the appropriate interest rate) for credit entering R02. Some credit may be used the entire year and the remainder for only a few months. The single sum capital restraint permits nothing more than a tenuous estimate of the appropriate C row entry on the borrowing activity.
3. The R02 entry in the B column is the quantity estimated by the procedure illustrated in Table 9.1. It represents the amount the operator has available in addition to those fixed or other capital items not considered in the model.
4. The R10 entry in the B column is an estimate of the amount of production credit available for the uses not circled on the coefficient side of the table. In this example we have assumed that no credit was being used. However, if the operator were already borrowing $1,000 against uncircled items in B, the R02 entry should be reduced to $9,000 and the R10 entry increased to $6,000. In this way interest is charged as an expense for all borrowed capital.

CAPITAL COEFFICIENTS FOR LIVESTOCK FACILITIES

The most frequent situation encountered in planning livestock activities is one where livestock facilities are available, to a limit, for some activities but not for others. Facilities may be available for farrowing fifteen litters at a time but no more. There may be no facilities for dairy cows or feeder cattle. In forming capital coefficients for the last two activities, capital required for facilities should be reflected in the coefficients. At this stage, facilities capital is allocable among activities and is not fixed. The same can be said of building capital for expanding existing activities. If the farm has adequate facilities for fifteen litters but you wish to weigh the alternative of expanding beyond this level, capital coefficients should be prepared for that portion of the activity which represents an expansion. In this case, an activity which provides for the production of hogs with existing facilities can be formed. No capital coefficients for facilities are entered. The activity is bounded at an upper limit of fifteen. A second hog production activity is provided for expansion beyond fifteen litters. It has the same coefficients as the first except that the requirement for additional facilities is reflected in the capital coefficient.

MODEL 9.3: *July Capital Restraint with Provisions for Additional Hog Facilities*

Explanation

1. The activities are the same as those in the previous model except that P15 has been added.
2. P15 is an activity which provides for producing hogs beyond 15 litters per year.
3. It is assumed that adequate facilities are already available on the farm to produce up to 15 litters of pigs. The model also assumes that adequate facilities are available to carry on other livestock activities at the level at which they are likely to enter the plan.
4. An upper bound, using the bounding option, is placed on P07 at 15 litters. Any additional hogs produced in the plan must be a product of P15.
5. P15 has a capital coefficient of $282, which is $75 higher than the same coefficient for P07. The difference is the capital outlay per litter associated with expanding capacity beyond 15 litters per year.
6. The cost per unit of hog facilities added should include taxes, depreciation, and insurance. The cost for these items on facilities now on the farm have been incurred already and hence were not included in the C row.

Points to Observe

1. Provisions for expanding activities other than hogs also could be built into the model. The relevant question in each case is the

MODEL 9.3

Row Type		B	P01	P02	P03	P04	P05	P06	P07	P08	P09	P10	P11	P13	P14	P15
N	C		—30	—81	—89	—58	79	92	446	1.20	—4	—1.25	.67	—.08	—.04	438.50
L	R01	4,800	5	16	16	12	20	15	36	0	1	0				36
L	R02	10,000	26.50	56.40	55.40	30.40	220	78	207	0	3.74	0		—1		282
L	R03	300	1	4												
L	R04	36	0		4	4										
L	R05	0	—90	—176	—124	—66	5	60	210	1	0	—1				210
L	R06	0	0	—3.2	—1.8	—3.6	4	1	1	0	1	0				1
L	R07	0	0	0	0		2	1	0	0	—1	0				
L	R08			—56	—46	—46							1			
L	R09	0		0				150							—1	
L	R10	5,000												1		
	Upper Bound								15							
	Lower Bound															

capacity of existing facilities. The model should be constructed so that part of the activity which represents an extension beyond existing facilities includes an appropriate capital coefficient for facilities.

2. There may be no facilities for certain livestock activities on the farm. In this case the capital coefficient (and the C row) would include provisions for facilities throughout the entire range of the activity. Separate expansion activities are not appropriate since the basic activity through all levels includes a capital coefficient for facilities.
3. Restraints in this model are imposed by two methods—the explicit B column restraint and the use of the bounding option.
4. The bounding option provides a simple device for restraining activities to the appropriate level in many situations. Where several activities are subject to the same facility restraint, an additional facility transfer row and facilities accounting activity may be necessary. The facilities activity can be bounded, thus restraining the total of all activities using the facility to the capacity available. Such a situation arises, for example, where two or more hog activities compete for hog facilities or two or more cattle feeding activities compete for the same feedlot and housing capacity.

DEFINITION OF THE PLANNING PERIOD

Introducing capital restraints into models requires definition of the length and nature of the period involved in the planning process because of the growth in capital investment typical of a successful farm operation.

Thus far in our examples and in our discussion of capital treatment we have been vague about the planning period to which our models applied. Any further treatment of the capital problem requires that we pause to clarify the conceptual difficulties surrounding the definition of the planning period.

Although farm production in the Corn Belt tends to be a seasonal activity with most operations carried out each year, it is not easy to define a beginning or an end to the production cycle. Farm businesses are organized around a number of activities, each with its own cycle. Furthermore, the timing of cycles from one activity to another is usually different; they begin on different dates and continue for different time periods. Therefore it is impossible to select any date on which to begin the planning year without encountering activities that are already in process. Figure 9.1 has been constructed using selected

	P01 Continuous Corn; Winterset Silty Loam	P02 CCOM; Winterset Silty Loam				P05 Beef Cow - Calf Calf Sold	P06 Feeding Yearling Steers	P07 Sow and Two Litters	
Jan						Cows		Winter Gilts	Jan
Feb			Prepare & plant oats w/ seeding					Sp. pigs born	Feb
March	Prepare seedbed; plant			Prepare & plant 1st year corn	Prepare & plant 2nd year corn	Calves born		Sows farrow; Fall pigs sold	March
April									April
May		Harvest and/or pasture meadow						Sows bred	May
June	Cultivate			Cultivate	Cultivate	Cows bred	Steers sold		June
July			Harvest oats						July
Aug								Spring pigs sold	Aug
Sept							Calves bought	Sows farrow; Fall pigs born	Sept
Oct	Harvest	Plow		Harvest	Harvest	Calves sold		Gilts saved	Oct
Nov	Plow			Plow				Sows sold; Gilts bred	Nov
Dec									Dec

	P01	P02				P05	P06	P07	
Jan					Prepare & plant oats w/ seeding			Winter Gilts	Jan
Feb								Sp. pigs born	Feb
March	Prepare seedbed; plant	Prepare & plant 1st year corn		Prepare & plant 2nd year corn		Calves born		Sows farrow; Fall pigs sold	March
April									April
May			Harvest hay and/or pasture meadow	Cultivate				Sows bred	May
June	Cultivate	Cultivate					Steers sold		June
July					Harvest	Cows bred			July
Aug								Spring pigs sold	Aug
Sept			Fall plow				Calves bought	Sows farrow; Fall pigs born	Sept
Oct	Harvest	Harvest		Harvest		Calves sold		Gilts saved	Oct
Nov	Plow	Plow						Sows sold; Gilts bred	Nov
Dec									Dec

	P01	P02				P05	P06	P07	
Jan				Prepare & plant oats w/ seeding				Winter Gilts	Jan
Feb								Sp. pigs born	Feb
March	Prepare seedbed; plant	Prepare & plant 2nd year corn	Prepare & plant 1st year corn			Calves born		Sows farrow; Fall pigs sold	March
April									April
May					Harvest hay and/or pasture meadow			Sows bred	May
June	Cultivate	Cultivate	Cultivate				Steers sold		June
July				Harvest oats		Cows bred			July
Aug								Spring pigs sold	Aug
Sept							Calves bought	Sows farrow; Fall pigs born	Sept
Oct	Harvest					Calves sold		Gilts saved	Oct
Nov	Plow	Harvest	Harvest		Plow			Sows sold; Gilts bred	Nov
Dec			Plow						Dec

FIG. 9.1. Illustration of activity cycles, Winterset Example.

activities from the Winterset example to illustrate how the farm production process cycles from one production period to another. In this example we have tried to find neither a beginning nor an end to the process. Regardless of how far we extended the diagram in either direction, we could never find a time when there was no activity in process. Instead we have been content to arrange the diagram so that all the activities or parts of activities taking place during a year will be encompassed within the middle year of the diagram.

The twelve months within the area outlined by the heavy black lines is the time period we have focused on in the Winterset example. We have sought to define the set of activities which, given our re- restraints and price expectations, would maximize the value of the program for this period. Our planning model does not indicate the path to be followed in moving the farming operation from its present organization to the optimum plan. It assumes implicitly that the same pattern will be repeated year after year. But the plan cannot in reality remain optimum where capital limits the plan when the business is successful, because available capital will increase from one time period to another.

Two methods of incorporating capital growth into the farm planning process are discussed below.

RECURSIVE CAPITAL GROWTH MODELS

In this modification one develops a series of bench mark plans to point the direction in which the business might move as the anticipated capital accumulation progresses. Computationally it involves two or more stages. Net income and hence potential increase in net worth are estimated for the initial production period in the first optimization. The estimated growth in net worth in turn indicates by how much operator-supplied capital plus borrowing capacity might reasonably be expected to increase yearly. Then this information is used to modify the B column capital entry in a second optimization. From then on the capital restraint for each planning period is developed on the basis of the net income provided by the solution for the previous year. One can project plans into the future in this fashion for many years at relatively low cost. But because of the compounding of errors in predicting income and capital growth inherent in such a chain of solutions, bench mark plans projected for more than four or five years in the above manner are of doubtful value.

Restraints other than capital may also change. To the extent that the changes are predictable, they too may be incorporated into future plans by adjusting the appropriate entry in the B column.

YEARLY UPDATING OF PROJECTED PLANS

A plan prepared in 1974 and projected until 1977 involves predicting prices, yields, and resource availabilities three years in advance. Because of the uncertainty of such predictions, a plan so projected is not likely to represent the optimum course of action when 1977 actually arrives. Yet there are advantages in extending the planning horizon beyond a single year and updating this projection each year. More accurate information regarding price expectations, production coefficients, and the nature of resource restraints would be available on January 1, 1975, than on January 1, 1974. One realistic approach to farm planning is preparing a plan every year for the year immediately ahead and at the same time projecting the plan three or four years into the future, based on the best information currently available.

Certain costs difficult to incorporate into any planning model are incurred when an operator changes from one farm plan to another. Where yearly programming suggests major departures from past plans, the stability of the programmed solution should be subjected to close scrutiny. The output from the programming process provides one basis upon which to judge whether shifts are really desirable. Some shifts clearly are not reasonable and occur only because our model does not take into account all the costs and inflexibilities arising from abrupt changes in the business. The shadow prices of activities previously a part of the business but not entering the new solution should be examined. You will recall that shadow prices estimate the decrease in income that would result from forcing into the program one unit of activity which has been excluded from the programmed solution. Furthermore, the range of prices over which an activity would remain in the basis solution is important (see Chapter 7). If a major adjustment in the farming program is dependent on only a small change in the price of an activity, one should be wary of making the adjustment since all price expectations contain an element of uncertainty. Furthermore, in such situations the addition to income may be so small that it does not compensate for the hidden costs of a major change in farm organization.

In some planning situations it may be clear that major shifts from one activity to another are not feasible. Such changes may not be possible because of a lack of facilities or because the operator is not capable of executing quick and drastic shifts in the business. In these circumstances year-to-year change in the plan may be restricted by a system of flexibility restraints. Although they may be used with recursive step-by-step models, their use is illustrated in connection with multiyear models in the following section.

TREATMENT OF CAPITAL IN MULTIYEAR MODELS

Thus far we have focused attention on models that treat the phenomenon of capital accumulation in the simplest manner possible. We have done so at the expense of precision and assumptions which do not square fully with the facts in real planning situations. The multiyear model (also referred to as a dynamic model) discussed in this section treats capital restraints and money more realistically than any other approach. But it does so at the price of a much larger model and more work on the part of the planner.

The multiyear model defines restrictions, price expectations, production coefficients, and the range of alternative activities for a planning period of four or five years. A plan for each year emerges from a single optimization. Capital available during any production period is estimated within the model, thus resolving the dilemma wherein capital availability is a function of income and vice versa. Such models have the disadvantage of mushrooming into matrices of immense size and complexity when structured realistically. Readers who are troubled by the assumptions essential to less sophisticated models may find comfort in the more rigorous formulations available to those willing to invest more time in model building and to incur higher computational costs.

In the multiyear model each restraint appears in a separate row for each production period contained within the model. Thus in a four-year planning model where the restriction on land is 320 acres, the latter figure would be entered in the B column on four separate rows. Activities are likewise specified by time periods. Thus corn production t_1 (the first year) is an activity distinct from corn production in t_2 (the second year) and in t_3 and t_4. Provisions are made for transferring capital and feed from one period to another. Finally an activity to invest capital in an off-the-farm use at some realistic rate is included for each year of the planning period, thus specifying an off-farm opportunity cost that must be met before capital will be allocated to any on-farm activity. This opportunity cost for capital offsets differences among activities in the temporal pattern of their cost and return flows. In effect the activity for off-farm investment reflects the discount rate in its C row coefficient. The structure of the multiyear model is illustrated in Model 9.4 where a simplified version is presented.

MODEL 9.4: *Treatment of Capital in a Multiyear Model*

Explanation

This model is a highly simplified version which spans a total of four years. The size of model growing from the limited range of production alternatives considered illustrates the proliferation in activities and restraints characteristic of the multiyear planning approach.

Included in the model are flexibility restraints to prevent unrealistic year-to-year fluctuations in the level of activities. To simplify study of the flexibility restraint system, the restraints in this section of the model have been drawn together and presented in a subsequent section entitled Submodel 9.4.

Activities

P101, P201, P301, P401 are corn growing activities. P101 represents the first year, P201 the second, P301 the third, and P401 the fourth. In each case the unit of activity is one acre. In this set of activities the capital coefficients illustrate a situation where payment for fertilizer and seed is made upon delivery.

P102, P202, P302, P402 are corn growing activities. P102 represents the first year, P202 the second, P302 the third, and P402 the fourth. In each case the unit of activity is one acre. As contrasted from the activities above, this set reflects payment for seed and fertilizer after harvest.

P103, P203, P303, P403 are corn harvesting activities. P103 refers to the first year, P203 to the second, P303 to the third, and P403 to the fourth. The unit of activity is one acre.

P104, P204, P304, P404 are corn selling activities. P104 relates to year 1, P204 to year 2, P304 to year 3, and P404 to year 4. For each, the unit of activity is one bushel. Corn is sold in November.

P105, P205, P305 are corn selling activities. P105 refers to the first year, P205 to the second, P305 to the third. The activity unit for all three activities is one bushel. The activities represent selling in July.

P405 is a corn selling activity for the last (fourth) year. The unit of activity is one bushel. Corn is sold at the end of the planning period (December 31 of the fourth year). P405 may also be looked upon as an inventory accounting activity.

P106, P206, P306 are hog raising and selling activities. Each activity unit represents one sow and two litters, the first farrowed in July and the second in January. P106 represents a July litter in year 1 and a January litter in year 2, P206 refers to a July litter in year 2 and a January litter in year 3, P306 relates to a July litter in year 3 and a January litter in year 4.

P406 is a hog raising and selling activity during the fourth year of the planning period. It represents one litter farrowed in July and sold. Since the second litter would not arrive until January of the fifth year, the activity provides for rebreeding and maintaining the sow until the termination of the planning period, i.e., the end of the fourth year. The sow is inventoried at this time and the appropriate coefficient entered in the inventory row. Appropriate adjustments are also made in the C row to reflect the fact that only one litter is sold.

P107, P207, P307, P407 are activities which pay living expenses and property taxes. P107 refers to year 1, P207 to year 2, P307 to year 3, and P407 to year 4. The unit of activity is $6,800.

P108, P109, P110
P208, P209, P210
P308, J309, P310
P408, P409, P410 are activities which transfer capital between quarters within years. P108 transfers capital from the 1st to the 2nd quarter, P109 from the 2nd to the 3rd quarter, and P110 from the 3rd quarter to the 4th quarter in year 1. P208, P209, P210 perform the same function for year 2; P308, P309, P310 for year 3; and P408, P409, P410 for year 4. In each case the unit of activity is $1.00.

P111, P211, P311 are activities which transfer capital from the last quarter of one year to the first quarter of the next. P111 is an activity which transfers capital from the fourth quarter of year 1 to the first quarter of year 2, P211 serves the same function between years 2 and 3, and P311 between years 3 and 4. The activity unit in each case is $1.00.

P112, P113, P114, P115 are capital borrowing activities for the 1st, 2nd, 3rd, and 4th quarters of year 1.

P212, P213, P214, P215 are capital borrowing activities for the 1st, 2nd, 3rd, and 4th quarters of year 2.

P312, P313, P314, P315 are capital borrowing activities for each quarter of year 3.

P411, P412, P413, P414 are capital borrowing activities for each quarter of year 4.

In all the activities above each unit borrowed represents $1.00 for a period of three months.

P116, P117, P118, P119
P216, P217, P218, P219
P316, P317, P318, P319
P415, P416, P417, P418 are capital lending activities. They permit capital to be invested off the farm and hence represent an opportunity cost for farm-invested capital. Each activity corre-

sponds to one quarter of the 16-quarter planning period. The unit of activity is $1.00.

P419 is an activity to inventory capital at the end of the planning period.

Restraints

R101, R201, R301, R401 are restraints on the number of tillable acres which may be cropped. R101 refers to year 1, R201 to year 2, R301 to year 3, and R401 to year 4. Each B column unit represents one acre.

R102, R202, R302, R402 are standing corn transfer rows. R102 transfers standing corn during year 1, R202 during year 2, R302 during year 3, and R402 during year 4. The transfer unit is one acre.

R103, R203, R303, R403 are corn grain transfer rows. R103 transfers corn harvested in year 1, R203 in year 2, R303 in year 3, and R403 in year 4. The unit of transfer is one bushel.

R104, R105, R106, R107
R204, R205, R206, R207
R304, R305, R306, R307
R404, R405, R406, R407 are labor restraints. R104 restrains labor during the period from Jan. 1 to Mar. 31 of the first year, R105 from Apr. 1 to June 30 of the first year, R106 from July 1 to Sept. 30 of the first year, and R107 from Oct. 1 to Dec. 31 of the first year. R204–R207 perform a similar function for year 2, R304–R307 for year 3, and R404–R407 for year 4. For all the restraints above, each unit in the B column represents one hour.

R108, R109, R110, R111
R208, R209, R210, R211
R308, R309, R310, R311
R408, R409, R410, R411 are capital restraints. R108 restrains capital during the period from Jan. 1 to Mar. 31, R109 from Apr. 1 to June 30, R110 from July 1 to Sept. 30, and R111 from Oct. 1 to Dec. 31 of year 1. R208–R211 perform the same function for year 2, R308–R311 for year 3, and R408–R411 for year 4. Each B column unit in the case of all capital restraints represents $1.00.

R412 is an equality restraint which transfers capital accumulated at the end of the planning period to a capital inventorying activity. The transfer unit is dollars.

R001–R006 are flexibility restraints. To simplify discussion of their features, all the restraints which constitute the flexibility restraint system are condensed and presented later in Submodel 9.4.

UP BND1 supplies a series of bounds on borrowing activities which have the effect of restraining each borrowing activity to $10,000.

Points to Observe

1. We specify a definite starting and terminating date for the planning period. This enables us to be specific about the capital entries in the B column.
2. Activities for off-the-farm investment of capital are provided. This has the effect of discounting future returns so that an activity which produces income in a nearby time period (other things equal) is preferred to one producing more distant income.
3. Income is maximized over the entire period. Thus resources are used in t_1 (the first year) in such a way as to make a maximum contribution not only to t_1 but to t_2 and subsequent years as well.
4. The last year, t_4, ends with a series of activities that are terminated by liquidating the production in process at inventory rather than sale prices.
5. Transfer activities are provided whereby capital supplies are transferred from one period to another within the year and from one year to another. The planning process begins with $6,000 (R108, B); the $6,000 may be increased (or decreased) by the difference between income earned and family living and taxes (P107, P207, P307, P407) as it flows through the model.
6. The capital coefficients are not cumulative. On the contrary, only the capital that need be invested during a time period enters a coefficient on the appropriate capital row.
7. Fixed cost and family living activities are provided. Their level is predetermined and restrained to a specific quantity during each year by use of a lower bound.
8. All restraints are repeated for each of the four years involved. Similarly, all activities are repeated four times except the transfer activities from the fourth year onward. The capital for the fourth year would typically be transferred to a capital accounting activity via a capital transfer row.
9. Capital borrowing activities (P112, P113, P114, and P115) and their counterparts in subsequent years are restrained by the UP BND1 to $10,000. Although $10,000 capital can be borrowed during each quarter of the year, each of the activities also provides for repaying the loan with interest in the subsequent period. Thus the maximum size of the loan that can be outstanding at any one time is $10,000.
10. The coincidence of the time periods of the four capital restraints and the off-farm investment activities permits competition between the capital transfer activities and the investment activities.

MODEL 9.4

		B	P101	P102	P103	P104	P105	P106	P107	P108	P109	P110	P111	P112	P113	P114	P115	P116	P117	P118
	C		-35.01	-35.54	-10.	.92	1.06	494.92	-6800.					-.021	-.021	-.021	-.021	+ .015	.015	.015
	R101	400	1	+1																
	R102		-1	-1	+1															
	R103				-100	1.0	1.0	210												
	R104	640						1.0												
	R105	720	3.0	3.0				1.0												
	R106	680	.2	.2				7.5												
	R107	720	.3	.3	1.1			6.5												
	R108	6,000	.25	.25				5.80	2,200	1				-1				1		
	R109		34.35	7.85				3.95	1,200	-1	1			1.021	-1			-1.015	1	
	R110							-68.01	2,200		-1	1			1.021	-1			-1.015	1
	R111		.41	27.44	10.0	-.92		-218.75	1,200			-1	1			1.021	-1			-1.015
	R201	400																		
	R202																			
	R203																			
	R204	640						7.0												
	R205	720						5.0												
	R206	680																		
	R207	720																		
	R208							-56.73					-1				1.021			
	R209							-161.18												
	R210						-1.06													
	R211																			
	R301	400																		
	R302																			
	R303																			
	R304	640																		
	R305	720																		
	R306	680																		
	R307	720																		
	R308																			
	R309																			
	R310																			
	R311																			
	R401	400																		
	R402																			
	R403																			
	R404	640																		
	R405	720																		
	R406	680																		
	R407	720																		
	R408																			
	R409																			
	R410																			
	R411																			
	R412																			
	R001		+1	1																
	R002																			
	R003	10						+1												
	R004	20						-1												
	R005	30																		
	R006	30																		
UP	BND1													10,000	10,000	10,000	10,000			
LØ	BND1								1											

P119	P201	P202	P203	P204	P205	P206	P207	P208	P209	P210	P211	P212	P213	P214	P215	P216	P217	P218	P219
.015	-35.01	-35.54	-10.0	.92	1.06	402.92	-6800.					-.021	-.021	-.021	-.021	.015	.015	.015	.015
1																			
	1	1																	
	-1	-1	1																
			-100	1	1	210													
						1.0													
	3.0	3.0				1.0													
	.2	.2				7.5													
	.3	.3				6.5													
-1.015	.25	.25				5.80	2200	1				-1				1			
	34.35	7.85				3.95	1200	-1	1			1.021	-1			-1.015	1		
						-68.01	2200		-1	1			1.021	-1			-1.015	1	
	.41	27.44	10.0	-.92		-218.75	1200			-1	1			1.021	-1			-1.015	1
						7.0													
						5.0													
						-56.73					-1				1.021				-1.015
						-161.18													
					-1.06														
	-1	-1																	
	+1	+1																	
						-1													
						+1													
												10000	10000	10000	10000				
							1.												

MODEL 9.4 (continued)

P301	P302	P303	P304	P305	P306	P307	P308	P309	P310	P311	P312	P313	P314	P315	P316	P317	P318	P319	P401
-35.01	-35.54	-10.0	.92	1.306	494.92	-6800.					-.021	-.021	-.021	-.021	.015	.015	.015	.015	-35.01
1	1																		
-1	-1	1																	
		-100	1	1	210														
					1.0														
3.0	3.0				1.0														
.2	.2				7.5														
.3	.3				6.5														
.25	.25				5.80	2200	1				-1				+1				
34.35	7.85				3.95	1200	-1	1			1.021	-1			-1.015	1			
					-68.01	2200		-1	1			1.021	-1			-1.015	1		
.41	27.44	10.0	-.92		-218.75	1200			-1	1			1.021	-1			-1.015	1	
																			1
																			-1
					7.0														
					5.0														3.0
																			.2
																			.3
					-56.73					-1				1.021				-1.015	.25
					-161.18														34.35
				-1.06															
																			.41
-1	-1																		
					1														
					-1														
											10000	10000	10000	10000					

P402	P403	P404	P405	P406	P407	P408	P409	P410	P411	P412	P413	P414	P415	P416	P417	P418	P419	
-35.54	-10.0	.92	.94	373.61	-6800				-.021	-.021	-.021	-.021	.015	.015	.015	.015		C
																		R101
																		R102
																		R103
																		R104
																		R105
																		R106
																		R107
																		R108
																		R109
																		R110
																		R111
																		R201
																		R202
																		R203
																		R204
																		R205
																		R206
																		R207
																		R208
																		R209
																		R210
																		R211
																		R301
																		R302
																		R303
																		R304
																		R305
																		R306
																		R307
																		R308
																		R309
																		R310
																		R311
1																		R401
-1	1																	R402
	-100	1	1	130.0														R403
				1.0														R404
3.0				1.0														R405
.2				7.5														R406
.3				6.5														R407
.25				5.80	2200	1			-1				1					R408
7.85				3.95	1200	-1	1		1.021	-1			-1.015	1				R409
				-68.01	2200		-1	1		1.021	-1			-1.015	1			R410
27.44	10.0	-.92		-218.75	1200			-1			1.021	-1			-1.015	1		R411
			-.94	-85.00								1.021				-1.015	+1	R412
																		R001
																		R002
																		R003
																		R004
				-1														R005
				+1														R006
									10000	10000	10000	10000						BND1

The latter outcompete the transfer activities, and as a result no funds are transferred to latter periods via the transfer activities. This result is an artificial one growing out of the treatment of time as discrete rather than continuous. It would not arise if the lending periods were of longer duration than the capital restraint periods.

Instead of four restraints per year a more realistic model would include six restraints (2-month periods), twelve restraints (1-month periods), or even more. With a greater number of restraints the length of time a capital lending activity would span typically would not coincide with the duration of the restraints. Thus an activity which invested capital for 90 days would reflect the demand made on capital via a (+1) in restraint R_t and the yield from the investment by a (—1.015) in restraint R_{t+2}. Alternatively, in the model above, extending the length of off-farm investment periods to six months would reduce the distortion and eliminate the redundancy of activities, since the minimum off-farm investment period would be longer than the capital restraint period.

SUBMODEL 9.4: *Flexibility Restraints*

Explanation

Restraints R001 through R006 impose limits on the extent to which certain of the activities can change from one year to another. As suggested earlier, some costs associated with major shifts in activities typically are not reflected in the coefficients. It is possible to define a wide range of flexibility restraints. At one limit activities could be restrained in such a way that the activities mix could not change during the planning period. In this case there is little point in constructing an elaborate multiyear model; one should settle for a less complex formulation where a single activity spans all the years in the planning period. At the other extreme the flexibility restraints can be defined so loosely that they have no bearing on the optimization process and hence are redundant. There are few guidelines to which the planner can turn in deciding how much year-to-year change to permit. He must resort to his own intuition as to the amount of shifting about that can occur without doing violence to the assumptions on which the coefficients in the model are based.

Activities

All the activities listed below are from Model 9.4 and are the only activities in that model directly affected by the flexibility restraints.

P101 = first-year corn production; seed and fertilizer paid for upon delivery
P102 = first-year corn production; seed and fertilizer paid for after harvest
P106 = hogs farrowed in year 1 and year 2
P201, P202 = corn production in year 2
P206 = hogs farrowed in year 2 and year 3
P301, P302 = corn production in year 3
P306 = hogs farrowed in years 3 and 4
P406 = hogs farrowed in year 4

Restraints

R001 = a restraint which forces the number of acres of corn grown in year 2 to be equal to or exceed the number of acres of corn in year 1. The unit of restraint is one acre.
R002 = a restraint which forces the number of acres of corn grown in year 3 to be equal to or exceed the number of acres of corn in year 2.
R003 = a restraint to insure that the number of sows farrowed in year 2 is equal to or greater than 10 less than the number farrowed in year 1.
R004 = a restraint to insure that the number of sows farrowed in year 2 does not exceed by more than 20 the number farrowed in year 1.
R005 = a restraint which permits the number of sows farrowed in the 4th year to be no more than 30 less than the number farrowed in the 3rd year.
R006 = a restraint which permits the number of sows farrowed in the 4th year to exceed the number farrowed in the 3rd year by no more than 30.

Points to Observe

1. Note that the sum of P201 and P202 (corn, year 2) must be equal to or greater than the sum of P101 and P102 (corn, year 1). When the −1's of R001, P201, and P202 are transposed to the B column side of the equation, they become positive. As a result, capacity for P101 and P102 to enter the plan is generated.
2. P201 and P202 are linked to P301 and P302 in the same way. Thus acreage of corn in year 3 $\geq$ acreage of corn in year 2; acreage of corn in year 2 $\geq$ acreage of corn in year 1. Then acreage of corn in year 3 $\geq$ acreage of corn in year 1.

	B	Year 1 Corn P101	Year 1 Corn P102	Years 1 & 2 Hogs P106	Year 2 Corn P201	Year 2 Corn P202	Years 2 & 3 Hogs P206	Year 3 Corn P301	Year 4 Corn P302	Years 3 & 4 Hogs P306	Year 4 Hogs P406
R001	0	1	1		—1	—1					
R002	0				1	1		—1	—1		
R003	10			1			—1				
R004	20			—1			1				
R005	30									1	—1
R006	30									—1	1

3. Flexibility to *decrease* or *increase* hog production from year 1 to year 2 is provided by R003 and R004. R003 permits the number to decrease from year 1 to year 2 but restricts the decrease to 10 units. Thus the number farrowed in year 1 $\geq$ the number farrowed in year 2 plus 10; or number farrowed in year 2 $\geq$ number farrowed in year 1 minus 10. R004 provides that farrowings in year 2 may exceed those in year 1 by 20. The net result of the two restraints is that hog production in year 2 may flex over a range of 10 less than to 20 more than in year 1.
4. The change in farrowings from year 2 to year 3 is not restricted in any way by the system of flexibility restraints.
5. Change from year 3 to year 4, however, is restrained by R005 and R006. R005 permits the number in year 4 to drop 30 below the number in year 3. R006 allows volume of farrowings to increase by 30. Thus farrowings in year 4 must equal those in year 3 plus or minus 30.
6. Variations in flexibility restraints are almost limitless. Production can be restrained from rising or declining over time. Obviously, the numbers in the B column can be altered to permit any level of absolute change.
7. Note that we have used only +1's or —1's in the flexibility matrix. Relative change could be controlled by appropriate alterations in the coefficients. For example, if the —1 at the intersection of R001 and P201 were changed to a —1.1, a decrease of $9\% \left(\frac{1.0}{1.1} \times 100 \right)$ would be permitted.

CHAPTER TEN

SIMULATION FROM PROGRAMMING MODELS

To approach the planning process as if a single optimum plan were the only objective ignores many of the benefits which can flow from developing comprehensive programming models. Once a model has been constructed and debugged, a process of experimentation can be undertaken to provide insight into a variety of relationships. This process is often referred to as simulation. So simple a modification as introducing different restraint levels via the multiple B column procedure can be viewed as a form of simulation. Likewise, altering price relationships through the use of multiple C rows is essentially an exercise in simulation, since one is experimenting with the effect price changes induce in the optimum mix of activities and level of income.

SIMULATING PAST PERFORMANCE

A useful first step in developing a planning model for an ongoing business is to attempt to restructure its existing organization. Such a model can be based on the past year or even the current year, if the farming program for the year is sufficiently well along to permit reasonable projection of its eventual outcome. In this application of the simulation process one attempts to describe, within a programming model framework, the relationships that have prevailed among activities, coefficients, and restraints. More specifically, the entire range of activities encompassed within the business—purchasing, selling, hiring, borrowing, and renting—must be defined and coefficients provided for each row. The production coefficients should estimate the input of each resource category per unit of activity actually achieved on the farm. The coefficients in the objective function (C row) should be based on prices actually received. Likewise, the B column entries should reflect the resource position of the business.

To avoid difficulties with infeasible solutions, provisions should be made to supplement resource quantities through input acquiring activities. This will not be necessary in the case of some restraints such as land and livestock facilities where the B column quantities are clearly defined and input coefficients for the relevant activities lend themselves to accurate specification. The next step is to bound each activity at the level at which it was conducted during the period being simulated.

Once the model has been processed, the result can be checked at several points to ascertain how well the simulated outcome conforms to what actually happened. The value of the program where fixed cost paying activities have been included should approximate the level of income actually achieved. Labor, including hired labor, required in the simulated operation should conform to the supply that was actually available. Likewise, feed grains and hay raised and purchased should correspond to quantities fed and sold.

Reasonable correspondence of the simulated outcome to what actually occurred provides an excellent foundation on which to structure planning models. The reader should note that the procedure just described concentrates on a carefully organized description of what *has happened* in the business as a step toward greater realism in planning models which look to the future.

MODIFYING SIMULATION MODELS FOR PLANNING PURPOSES

A model that realistically simulates the past operation of the business requires several modifications when used for planning purposes:

1. The range of activities included in the model should be extended to encompass other activities which the planner may wish to test. The activities added may provide for new crops or new classes of livestock; or they may be modifications of existing activities such as shifts in the timing of production and marketing, changes in methods of production, or shifts in the weights and/or grades of products produced. The addition of new activities to the model does not mean they will necessarily become a part of the plan finally chosen. The new activities may be left unbounded to compete for a role in the plan, or they may be bounded at nonzero in one or more of the fixed bound sets and hence forced into one or more alternative plans being tested.
2. The B column quantities should be updated to conform to

the expected resource position of the business during the planning period. As an example, where land has been added to the farm, the B column coefficient in the appropriate land row or rows must be increased.

3. Prices of products and inputs in the simulation model reflect past price relationships; those in the planning model should reflect expected prices.
4. Production coefficients such as bushels of corn per unit of land, quantities of feed required per unit of animal production, and labor required per unit of activity should be adjusted to reflect expectations rather than past performance. Such adjustments are especially important where conditions during the simulated period were unusual.

SIMULATION OF INCOME VARIABILITY

The emphasis in programming resulting thus far has centered on the optimum mix of activities, the shadow prices, and the level of income attained. The models presented have been based on single value yield and price expectations with the exception of multiple C rows and parametric price programming. Thus the extent to which yield variance and price fluctuations may be expected to induce variability into the income flow has largely been ignored. Yet some farm operators may be almost as interested in the range within which income can be expected to fluctuate as they are in an estimate of the most likely income from a plan.

In reality we can predict neither the average income nor its variability with certainty, because we are always confronted with uncertain yield and price expectations. Typically we turn to the record of the past in attempting to formulate price and yield coefficients. As discussed before, we focus on the averages over some recent period with appropriate adjustments for any obvious trends. Just as we look to the past as a foundation for predicting the most likely yields and prices and hence income, we can also look to the past to provide insight into the fluctuations we can expect in the future.

For example, we can begin by developing price and yield coefficients in the usual way, construct a model, and optimize. Next, in order to provide insight into the income fluctuations one might anticipate if such a farming program were actually followed, the yield history on a farm of similar type can be established. Likewise, one can reconstruct the farm price pattern for products of similar quality to those anticipated under the plan and marketed in a seasonal pattern similar to that projected in the plan. Armed with historical price and yield patterns for (say) ten years, one can then study the impact of yield and price changes on income variation by altering the model already optimized.

The key activities that determine the shape of the program should be bounded at the optimized level. If this is not done, one will find himself studying both the variation in income and the variation in the optimum activity mix which result from yield and price fluctuations. Under certain circumstances one may be interested in studying both, but this implies that an operator can and is willing to make major shifts in his program from year to year in response to short-run shifts in his yield and price expectations. Although the typical operator undoubtedly can and does make minor year-to-year adjustments in his program as his expectations change, it is more realistic and useful to concentrate on the shifts in income that can be expected to occur under a stable farming program.

Where crops produced in the plan are used as feed for livestock, it may be necessary to add purchasing activities to assure that infeasibilities are not induced by downward fluctuations in crop yields.

Such an exercise in simulation as proposed above at first may seem forbidding in view of the fact that eleven optimizations would be involved. Accumulating the necessary price and yield coefficients may be burdensome if appropriate records are not close at hand. However, constructing a model to accommodate the changes in coefficients and subsequent optimizations need not be troublesome. The series of price coefficients can be accommodated readily by the multiple C routine which has already been described. The procedures for modifying coefficients discussed in Chapter 8 can be applied to successive adjustments in yield coefficients.

CHAPTER ELEVEN

MACHINE BUDGETING AND PARTIAL OPTIMIZATION

Thus far our discussions of farm planning have focused on optimizing, i.e., selecting the most advantageous mix of activities within a specified set of restraints. The reader also has been introduced to the concept of simulation in the previous chapter. Both in optimizing and in simulating, it has frequently been necessary to impose maximum, minimum, and equality restraints on activity levels. One method of restraining activity levels was illustrated in Model 3.12. A second and less troublesome method, the bounding option, was introduced in Model 3.13. The ease with which one can fix the level of one or a set of activities and obtain multiple solutions in a single computer run permits great flexibility in the application of the programming approach we have been studying. At one extreme we can use the model as little more than a convenient framework for computerizing the arithmetic involved in budgeting. At this extreme we fix the levels of the key activities in the model (i.e., develop a plan) and seek to determine the level of income the plan can be expected to generate and the quantity of each resource it would require. Alternatively, we seek to compare the projected outcomes from five or ten different activity mixes or alternative farm plans. We refer to this approach in subsequent discussions as computerized budgeting.

COMPUTERIZED BUDGETING AND PROGRAMMING: A COMPARISON

The computerized budgeting and the programming approach both have the same objective: to develop a farm plan that achieves as fully as possible the objectives of the business. The budgeting approach defines one or more specific alternatives to the existing organization and estimates their probable effect on resource requirements and income. At the other extreme the programming approach specifies the restraints impinging on the business, sets forth a wide range of alternative production activities, and then attempts to define the combination of activities that will maximize income.

The latter approach clearly is more comprehensive and hence more appealing conceptually since it seeks the one best solution. Although budgeting is essentially a trial-and-error procedure, often it is accepted more readily by the farm operator because it more nearly approximates his approach to decision making. He is managing a going business and seeks to change it from what it is to something better. This leads him to emphasize in his planning activities the effects of marginal rather than sweeping changes in the farm business.

PARTIAL OPTIMIZATION MODELS

Thus far we have introduced the two extremes: (1) computerized budgeting and (2) conventional programming. Between them lies a continuum wherein the two approaches may be combined in any degree desired. Thus we may fix the level of one or more activities and permit the level of the remaining activities to be optimized in the programming process. We shall refer to a planning model where part of the activity levels are fixed and others are optimized as a partial optimization model. To illustrate the application of the partial optimization technique, we refer again to the Winterset model from Chapter 4. In that conventional programming model we defined a set of restraints, coefficients, and activities and sought to determine the mix of activities which would maximize the value of the program. By contrast, in the partial optimization approach we fix the level of certain key activities and observe the effect on income and resource requirements.

MODEL 11.1: *Partial Optimization*

Explanation

1. The first 11 activities and all of the restraints are the same as the original Winterset example from Chapter 4. They are summarized below for convenience.

   P01 = growing and harvesting continuous corn. The activity unit is one acre.

   P02 = growing and harvesting corn and oats and growing meadow in a CCOM sequence on Winterset silt loam. The unit of activity is two acres of corn, one of oats, and one of meadow.

P03 = growing and harvesting corn and oats and growing meadow in a CCOM sequence on Shelby loam. The activity unit is two acres of corn, one of oats, and one of meadow.

P04 = growing and harvesting corn and oats and growing meadow in a COMM sequence on Shelby loam. The activity unit is one acre of corn, one of oats, and two of meadow.

P05 = raising and selling beef calves. The unit of activity is one beef cow.

P06 = feeding yearling steers. The unit of activity is one steer.

P07 = hog raising and selling. The unit of activity is one sow and two litters.

P08 = corn selling. The unit of activity is one bushel.

P09 = hay harvesting. The unit of activity is one ton.

P10 = corn buying. The unit of activity is one bushel.

P11 = oats selling. The unit of activity is one bushel.

R01 = a labor restraint. The B column entry is hours of labor.

R02 = a capital restraint. The B column entry is dollars.

R03 = a land restraint on Winterset silt loam. The B column entry is acres.

R04 = a land restraint on Shelby loam. The B column entry is acres.

R05 = a grain transfer row. The transfer unit is bushels of corn.

R06 = a standing meadow transfer row. The unit of transfer is tons.

R07 = a hay transfer row. The unit of transfer is tons.

R08 = an oats transfer row. The unit of transfer is bushels.

2. The following activities have been added:

P12 = a labor hiring activity. The unit of activity is one hour.

P13 = a capital borrowing activity. The unit of activity is one dollar.

P14 = a hay buying activity. The unit of activity is one ton.

These activities permit additions to the original B column quantities of labor and capital and to the supply of hay generated within the model. Activities to supplement production inputs are an essential feature of partial optimization models. In their absence the planner risks specification of activity mixes that are infeasible.

3. Three sets of fixed bounds for the livestock activities labeled Set A, Set B, and Set C have been defined as follows:

	P05	P06	P07
Set A	10	0	20
Set B	0	60	30
Set C	20	30	10

No bounds have been specified for any of the cropping activities. Hence they will be optimized, subject to the restraints specified, including the levels of livestock activities which have been dictated above.

4. Each of the three sets of activity bounds constitutes a separate optimization and results in a unique farm plan. One may specify as many sets as he wishes and arrange to have them processed in a single computer run.

Points to Observe

1. Three optimizations result from processing the model. One solution relates to each bound set. The three activities that have been bounded enter the solution in each of the sets at the level at which they were fixed. The other activities enter the plan to maximize income, given the conditions that the levels of three activities have been predetermined. The plans which result from each bound set are summarized below.
2. The quantities of resources used in each of the plans also may be deduced from the output. Doing so requires inspection of (a) the original B column coefficients, (b) the level of disposal activities (resources unused), and (c) the level at which activities designed to supplement resources entered the solution. In this example our interest focuses on labor (R01) and capital (R02), the labor hiring activity (P12), and the capital borrowing activity (P13). The plan for Set B uses both labor and capital in addition to the quantities originally contained in the B column. None of the plans involves purchase of corn (P10) but all include purchase of hay (P14).

	Set A	Set B	Set C	Original Model with No Activity Bounds
Value of the Program	14,169	13,808	9,854	19,417
P01	95.25	57.75	20.25	113.57
P02	13.69	23.06	32.44	9.11
P03	0	0	4.5	0
P04	4.50	4.50	4.50	4.5
P05	10.00	0	20.00	0
P06	0	60.00	30.00	0
P07	20.00	30.00	10.00	45.34
P08	7,028.50	0	3,828.50	2,598.63
P09	0	0	0	0
P10	0	346.50	0	0
P11	973.50	1,498.50	2,023.50	717.04
P12	0	291.75	0	0
P13	0	2,339.60	0	0
P14	20.00	60.00	70.00	0

MODEL 11.1

Row Type		B	P01	P02	P03	P04	P05	P06	P07	P08	P09	P10	P11	P12	P13	P14
N	C		—30	—81	—89	—58	79	92	446	1.20	—4	—1.25	.67	—1.50	—.07	—20.00
L	R01	2,400	5	16	16	12	20	15	36		1			—1		
L	R02	13,000	20	64	72	60	142	160	90		3	.4			—1	
L	R03	150	1	4												
L	R04	18			4	4										
L	R05	0	—90	—176	—124	—66	5	60	210	1		—1				
L	R06	0		—3.2	—1.8	—3.6	4	1	1		+1					
L	R07	0					2	1			—1					—1
L	R08	0		—56	—46	—46							1			
BOUNDS																
FX BND A							10	0	20							
FX BND B							0	60	30							
FX BND C							20	30	10							

	Original B Column Quantity	Amount in Disposal	Hired, Borrowed, or Rented	Net Quantity Used in Plan
Set A				
Labor (R01)	2,400 hours	730.75	0	1,669.25 hours
Capital (R02)	$13,000	$6,729.00	0	$ 6,271.00
Set B				
Labor (R01)	2,400 hours	0	291.75	2,691.75 hours
Capital (R02)	$13,000	0	$2,339.60	$15,339.60
Set C				
Labor (R01)	2,400 hours	515.75	0	1,884.25 hours
Capital (R02)	$13,000	$1,709.00	0	$11,291.00

3. From the example in Chapter 4, we know that optimizing the model resulted in a value of the program of $19,417. In the optimum plan, continuous corn (P01) and hog production (P07) were emphasized. None of the three livestock plans which were arbitrarily selected in establishing the three sets of fixed bounds is as profitable as concentrating on hog production alone. The levels of crop production activities for each of the three sets of livestock bounds differ from those in the fully optimized model but are optimum given the mix of livestock activities we forced into the plans. The differences in outcomes, although somewhat exaggerated, emphasize that the value of the program, except for a rare stroke of luck, will be lower for partially than for fully optimized plans. In no case would a partially optimized plan have a higher value of the program than one that had been fully optimized.
4. Shadow prices on the activities restrained by the bound sets show the direction of the adjustments in the plan needed to move toward an optimum. To illustrate, the shadow prices (income penalties) in Set A where P05 (beef cows) was fixed at 10, P06 (yearlings) at zero, and P07 (hogs) at 20 were as follows:

P05	−147.35
P06	− 45.09
P07	+148.91

Adding a unit of beef cows would reduce the value of the program by $147.35, and forcing in a unit of yearling feeding steers would result in a reduction of $45.09. On the other hand, adding a unit of hogs beyond the 20-unit limit would increase the value of the program by $148.91. Clearly, the shadow prices indicate that the optimum lies in the direction of more hogs and fewer beef cows and feeder cattle than this plan provides for.

PARTIAL OPTIMIZATION AND DECREASING COST ACTIVITIES

The problems arising from the assumption of linearity in the coefficients have already been discussed. To review, the difficulties become very real when one must specify a labor, capital, or cost coefficient for a livestock enterprise such as hogs with little basis for predicting the level at which the activity will enter the plan. Should labor requirements per litter of hogs be based on a 10-litter or a 40-litter operation? Or should the investment in facilities for dairy production be based on a 30-cow or a 50-cow herd? In both instances the magnitude of the coefficient selected depends heavily on the size of enterprise one anticipates entering the plan as he constructs the model. Although there is no fully satisfactory means of resolving the linearity dilemma, labor and capital coefficients can be treated more realistically within the partial optimization framework than with the conventional approach.

The partial model presented below illustrates the procedure followed in constructing a partial optimization model which takes into account decreasing input requirements per unit of output as the scale of the enterprise increases.

MODEL 11.2: *Decreasing Cost Enterprises*

Explanation

The model emphasizes a single enterprise—hog production—which is partitioned into three activities, each representing a unique level of production. Only two restraints, one on labor and the other on capital, are included in the illustration. It is assumed that the activities and restraints shown represent only a small segment of a much larger model.

P01 = a hog raising and selling activity. One unit of activity equals four litters farrowed January, April, July, and October. Coefficients relate to the activity entering the program at a level of 10.

P02 = same as above except coefficients relate to activity fixed at a level of 20.

P03 = same as above except coefficients relate to activity fixed at a level of 30.

R01 = a labor restraint. The B column units are hours.

R02 = a capital restraint. The B column units are dollars.

C = the objective function which is the net price, including a deduction for fixed cost, per unit of activity.

FX BND1, FX BND2, FX BND3, FX BND4 = fixed bound sets which specify the level at which P01, P02, and P03 will enter in each of four solutions.

Row Type		P01	P02	P03
N	C	1,000	1,030	1,050
L	R01	40	32	26
L	R02	330	260	210
Bounds				
FX BND1		0	0	0
FX BND2		10	0	0
FX BND3		0	20	0
FX BND4		0	0	30

Points to Observe

1. The coefficients for labor and capital have been estimated on the basis of the labor requirement anticipated for the level at which the activity has been fixed in the bounds section. Both labor and capital requirements per unit of activity decline as we move from P01 to P02 to P03.
2. Net prices increase as the scale of the enterprise expands because fixed costs per unit of activity decline. Fixed costs per unit of activity can be related to specific activity levels, i.e. 10, 20, 30 units.
3. To interpret the output one must compare the value of the program and resource requirements for four separate optimizations. The first plan will include no hogs, the second 10 units of hogs, the third 20 units, and the fourth 30 units.
4. P01, P02, and P03 will enter separate plans, but no combination of the three will enter the same plan because of the manner in which the activities have been bounded.
5. The output provides no information as to the outcome of a plan with 15 units of hog production. One could obtain information regarding in-between activity levels by proliferating the number of activities (and the number of solutions) relating to hogs.
6. The model could contain any number or mix of other activities which would be optimized around the bounds imposed on the hog activity.
7. More effort is involved in building models with elaborate decreasing cost relationships as suggested above, interpretation of the output is more laborious, and the plan which results typically will not be fully optimized. Yet where assumptions regarding labor and capital requirements are critical, the benefits from more realistic coefficients may outweigh the disadvantages.
8. Because of the very large number of plans which could result from

dissecting each enterprise by size into many activities, the procedure illustrated above lends itself best to a situation where one can reasonably limit the number of livestock enterprises to two or three. Cropping enterprises typically would not be dissected into activities on the basis of scale because the linearity assumption, although not fully realistic, is not unreasonable with respect to most cropping operations.

CHAPTER TWELVE

PROGRAMMING APPLICATIONS TO PROBLEMS OF MINIMIZATION

The programming illustrations encountered thus far have assumed maximization of the objective function. In the typical case we sought to maximize income subject to a set of restraints, usually resource limitations. However, certain types of problems in agriculture and elsewhere are more easily analyzed if we change the direction of the optimization and structure the model to *minimize* rather than maximize the objective function.

The maximizing framework presumes a mixture of resources fixed in proportion and quantity and a variety of potential alternative uses for them. The goal is specification of the combination of uses (activities) that will maximize income. By contrast, in the minimizing framework we specify a set of conditions in the form of maximum, minimum, and equality restraints which can be met by a variety of means (activities)—each with a cost attached—and seek the combination of means (activities) that permits fulfillment of the conditions at minimum cost.

This chapter contains six sections. In the first we introduce a minimization model on a level of simplicity comparable to the first maximization problem presented in Chapter 2. The next two sections describe models sufficiently realistic to introduce many of the difficulties one encounters in constructing models to solve real feed mixing problems. The fourth and fifth sections involve models to solve transportation and transshipment problems. The chapter concludes with an algebraic formulation of the minimization problem and a description of the application of the simplex method to its solution. The latter material, although not essential to constructing and optimizing models, is included for the student interested in a greater insight into what he is doing.

MODEL 12.1: *Minimizing Feed Costs*

Explanation

This example explores the problem of determining the least-cost method of providing winter feed for a herd of 25 beef cows during a period of 120 days. The problem has been simplified greatly to permit concentration on essential features of the minimization model.

1. The restraints in the model are as follows:

   R01 = a total protein restraint. The B column entry is pounds of protein.
   R02 = a total digestible nutrients (TDN) restraint. The B column entry is pounds of total digestible nutrients.
   R03 = a vitamin A restraint. The B column entry is thousands of International Units.

2. R01, R02, and R03 are minimum (greater than or equal to) restraints. R01 specifies that the 25 cows must be provided at least 4,500 pounds of protein but could be supplied more. A similar interpretation applies to the R02 and R03 restraints.
3. The activities are as follows:

   P01 = an activity supplying alfalfa hay. The activity unit is one pound.
   P02 = an activity supplying ground ear corn. The activity unit is one pound.
   P03 = an activity supplying soybean oil meal. The activity unit is one pound.

4. The coefficients for the activities appearing in R01, R02, and R03 represent the nutritional components one unit of each feed source would supply. Therefore, application of programming to feeding problems implies that the content of each potential ingredient in the ration is known.
5. The C row coefficients specify the cost of supplying one unit of each feed source (activity). In the minimization context, costs in the objective function carry a positive sign.
6. The signs of the other coefficients are also reversed from those in maximization models. Thus, although P01 supplies protein, TDN, and vitamin A, the coefficients in all three rows are positive.
7. Minimization models when fully specified require disposal

activities for minimum and maximum restraints and artificial activities for minimum and equality restraints. The row type designation causes appropriate disposal and artificial activities to be added in the computing routine.

Row Type		B	P01	P02	P03
N	C		.01	.016	.07
G	R01	4,500	.13	.074	.45
G	R02	27,000	.48	.73	.78
G	R03	54,000	2.16	0.0	0.0

Points to Observe

1. Although all restraints included in this illustration specify minimums, maximum and equality restraints may also be built into the model. If one wished to limit the total dry matter intake, a maximum restraint would be appropriate.
2. When optimized, the activities in the solution will indicate the optimum feed combination for wintering the 25 beef cows. The value of the program will indicate the minimum cost of feeding the herd given the nutrition requirements, the feed sources, and their analyses and prices specified in the model.

MODEL 12.2: *Least-Cost Ration Formulation*

Explanation

Although this model follows the pattern of the preceding example, the ration problem is formulated in greater detail and the solution is presented and its interpretation discussed. The problem defined in the model involves formulating the least-cost ration for wintering one 1,100-pound pregnant beef cow for one day.

1. The restraints, which specify the nutritional requirements to be met, are as follows:

R01 = minimum restraint on crude protein. The B column entry is pounds.
R02 = a minimum restraint on total digestible nutrients. The B column entry is pounds.
R03 = a minimum restraint on calcium. The B column entry is pounds.
R04 = a minimum restraint on phosphorus. The B column entry is pounds.

MODEL 12.2

Row Type		B2	B	P01	P02	P03	P04	P05	P06	P07	P08	P09	P10
N	C			.0015	.038	.012	.0025	.0036	.02	.0159	.07	.02	.086
G	R01	.8	1.5	0	2.62	.176	.051	.023	.089	.074	.458	.118	
G	R02	10	10.0	.46		.50	.48	.18	.80	.73	.78	.65	
G	R03	.044	.033	.0010		.0122	.0040	.0009	.0002	.0004	.0032	.0010	.21
G	R04	.037	.031	.0004		.0022	.0007	.0005	.0031	.0022	.0067	.0035	.14
G	R05	20,000	20,000			33,000	560	2,100	320				
G	R06	0		.93	—99	.90	.87	.29	.89	.86	.89	.89	0
L	R07		21.5	.93	1.0	.90	.87	.29	.89	.86	.89	.89	0

R05 = a minimum restraint on vitamin A. The B column entry is International Units.
R06 = a minimum restraint on the pounds of other feed that must be fed with each pound of urea in the ration. R06 may also be viewed as a transfer row, the role of which will be explained later.
R07 = a maximum restraint on the total intake of dry matter. The B column entry is pounds.

2. The activities are as follows:

P01 = an activity supplying ground corncobs. The activity unit is one pound.
P02 = an activity supplying urea. The activity unit is one pound.
P03 = an activity supplying high-quality alfalfa hay containing 90% dry matter. The activity unit is one pound.
P04 = an activity supplying corn stover containing 87% dry matter. The activity unit is one pound.
P05 = an activity supplying corn silage containing 29% dry matter. The activity unit is one pound.
P06 = an activity supplying No. 2 corn. The activity unit is one pound.
P07 = an activity supplying ground ear corn. The activity unit is one pound.
P08 = an activity supplying soybean oil meal. The activity unit is one pound.
P09 = an activity supplying oats. The activity unit is one pound.
P10 = an activity supplying mineral. The activity unit is one pound.

3. The C row gives the price (cost) per unit of activity. When the objective of the model is to minimize costs, coefficients in the C row are positive. This contrasts with the C row in the maximization model where coefficients designating a cost or negative income carry a minus sign.
4. The function of minimum restraint R06 requires explanation. P02 is an activity that supplies urea. The total ration cannot contain more than 1% urea on a dry matter basis. Hence, one unit of P02 entering the plan creates a requirement that 99 pounds of other material (on a dry matter basis) be included in the ration. Because P02 creates a requirement in R06, its coefficient in R06 is negative. Furthermore, the 99 pounds of dry matter from other sources is a minimum restraint. This means that a higher but not a lower proportion of other materials can be fed with the urea, or conversely urea may constitute less but not more than 1% of the ration. Thus each pound of urea (unit of P02) entering the solution has the effect of generating an entry of 99 in the B column of R06 as the coefficient (P02, R06) is transposed.

5. The salt requirement has not been included in this model but is assumed given free choice.

Interpretation of Output

1. The value of the program (.07016) represents the minimum cost at which the 1,100-pound beef cow could be fed for one day under the conditions specified in the original model.
2. The activities in the solution and their respective shadow prices are as follows:

		Activity Level	Reduced Cost
P01	(ground corncobs)	0	.00009
P02	(urea)	.155	0
P03	(alfalfa hay)	.257	0
P04	(corn stover)	20.566	0
P05	(corn silage)	0	.00209
P06	(corn)	0	.01471
P07	(ground ear corn)	0	.01162
P08	(soybean oil meal)	0	.05726
P09	(oats)	0	.01449
P10	(mineral)	.114	0

The activity levels represent the optimum mix of ingredients in the daily ration. Because in structuring the model all activity units were defined in pounds, activity levels in the solution are reported in pounds.

3. Activities in the solution at nonzero values have zero shadow prices; those reported at zero level (not in the final basis) have a positive shadow price. In maximization models the shadow prices on the activities represent income penalties; in the minimization framework they constitute cost penalties, because they specify by how much the cost of the ration would increase by forcing into the ration one unit of an ingredient that was available but was excluded in the process of optimization.
4. The status of the disposal activities and their shadow prices following optimization are as follows:

			Level of Slack Activity	Shadow Price
G	R01	(crude protein)	0	—.01450
G	R02	(TDN)	0	—.00254
G	R03	(calcium)	—.076	0
G	R04	(phosphorus)	0	—.61429
G	R05	(vitamin A)	0	—.00000
G	R06	(percent urea)	—2.785	0
L	R07	(total dry matter)	3.108	0

5. The level at which the disposal or slack activities enter the solution indicates which of the restraints were limiting. Where slack activity levels are greater than zero for a minimum restraint, increasing the level of the minimum restraint by the magnitude of its slack activity level would not increase the cost of the ration. More specifically, if minimum calcium requirements were increased by .076 units per day, the cost of the ration would not have changed. This is explained by the fact that the ration, once other requirements have been met, contains an abundance of calcium.
6. The shadow prices on the slack activities at zero level indicate by how much the cost of the ration would be reduced were the restraint relaxed by one unit. For example, reducing the minimum restraint on R01 (crude protein) by one pound would reduce the cost of the ration by $.0145.

MODEL 12.3: *Minimizing the Cost of Specified Feed Quantities*

Explanation

The two previous models have dealt with minimizing the cost of feeding beef cows during some period of time. In the following model the purpose is modified to determine the least-cost mix of ingredients in a given quantity of feed meeting a particular set of specifications. This is the problem confronting the feed mixer who sets out to mix (say) a ton of feed for a specific purpose. The example involves mixing one ton of growing ration for pigs 40–75 pounds in weight. Again in the interest of simplicity, the number of specifications (restraints) elaborated and the range of ingredients (activities) provided have been curtailed. However, the model may be extended to any number of restraints and activities desired by following the pattern illustrated.

1. The activities are as follows:

P01 = an activity supplying corn. The unit of activity is one bushel.
P02 = an activity supplying soybean oil meal. The unit of activity is 100 pounds.
P03 = an activity supplying oats. The unit of activity is one bushel.
P04 = an activity transferring cystine to permit up to 40% of the methionine requirement to be met by cystine. The unit of activity is one pound.

P05 = an activity supplying salt. The unit of activity is one pound.
P06 = an activity supplying a vitamin and mineral premix. The unit of activity is one pound.
P07 = an activity supplying a commercial source of methionine (D. L. Methionine). The unit of activity is one pound.
P08 = an activity supplying defluorinated phosphate. The unit of activity is one pound.

2. The restraints are as follows:

R01 = an equality restraint on the total quantity of ingredients. The B column entry is pounds.
R02 = a minimum restraint on the amount of crude protein. The B column entry is pounds.
R03 = a maximum restraint on fiber content. The B column entry is pounds.
R04 = a minimum restraint on lysine. The B column entry is pounds.
R05 = a minimum restraint on tryptophan. The B column entry is pounds.
R06 = a minimum restraint on methionine. The B column entry is pounds.
R07 = a minimum restraint on cystine. The B column entry is zero pounds because it is not an essential amino acid.
R08 = an equality restraint on salt. The B column entry is pounds.
R09 = a minimum restraint on calcium. The B column entry is pounds.
R10 = a minimum restraint on phosphorus. The B column entry is pounds.
R11 = a minimum restraint on niacin. The B column entry is milligrams.
R12 = a minimum restraint on riboflavin. The B column entry is milligrams.
R13 = a minimum restraint on pantothenic acid. The B column entry is milligrams.
R14 = a minimum restraint on vitamin B12. The unit is one milligram.

3. As in previous examples, the C row coefficients represent the cost per unit of the ingredients. The signs are positive because the model is designed to minimize the objective function.

Points to Observe

1. The total quantity of product is restrained to one ton (2,000 pounds) by use of an equality restraint.
2. The model contains all three restraint types.
3. In its application to a realistic feed mix problem, the analysis as presented would be extended in one or more ways. For example, the number of restraints on protein would be increased to assure

that a protein of minimum quality was provided. Restraints to force in minimum levels of other vitamins, minerals (both macro and trace), and other ingredients could also be added. Likewise, the range of materials from which the mix is formulated would be extended in a comprehensive application of the method.

4. Application of the method assumes that one is willing to define, without ambiguity, the specifications to be met by the final mix. Likewise, the model assumes that the programmer has knowledge of the ingredients supplied by each feed source. However, the need for knowledge, both as to specifications desired and the analysis of potential feed ingredients, is not unique to the method. Such knowledge is implied in feed formulation regardless of the analytical method applied.

Interpretation of Output

1. The least-cost mix of ingredients for one ton of feed meeting the specifications set forth in the B column is $59.83. Changes in specifications and/or relative prices of ingredients would change the cost. The reader should note also that the above estimate does not include the cost of mixing. A full-cost estimate would require a mixing activity forced in at a level of one ton or adding a mixing charge to the cost of the feed.

2. Ingredients that entered the solution are as follows:

Row Type		B	P01	P02	P03	P04	P05	P06	P07	P08
N	C		1.12	5.00	.70		.026	.086	.75	.057
E	R01	2,000	56	100	34		1	1	1	1
G	R02	320	4.98	43.8	4.08					
L	R03	100	1.12	6.0	3.74					
G	R04	14	.1008	2.7	.1224					
G	R05	2.6	.0504	.6	.0612					
G	R06	10	.0504	.8	.0612	1			1	
G	R07	0	.0504	.6	.0612	—1				
E	R08	10					1			
G	R09	13	.0112	.27	.034			.21		.33
G	R10	10	.0646	.21	.040			.14		.18
G	R11	12,727.3	0	1,381.81	244.18			400.		
G	R12	2,363.6	33.09	150	24.727			100.		
G	R13	10,000.	99.27	659.09	199.36			150.		
G	R14	10.000	0	0	0			.5		
UP	BND1					4				

P01 (corn)	18.80 (bushels)
P02 (soybean oil meal)	3.80 (100-pound units)
P03 (oats)	15.00 (bushels)
P05 (salt)	10.0 (pounds)
P06 (vitamin and mineral premix)	20.0 (pounds)
P07 (methionine)	1.1 (pounds)
P08 (defluorinated phosphate)	25.5 (pounds)

3. The activity P04 has a shadow price of —.737 because it was profitable to transfer more cystine (which was available) to methionine, but cystine can replace at most 40% of the methionine requirement if the latter is limiting. By bounding P04 at an upper limit of 4 (40% of 10) we set the maximum at 40% replacement.

   No nonzero shadow prices for activities appear in the output except P04, because all other activities in the original model are in the basis (have a nonzero value in the solution).
4. The report in the row section is as follows:

	Slack Activity		Level	Shadow Price
E	R01	(weight)	0.0	—.0177
G	R02	(protein)	—1.43	0.0
L	R03	(fiber)	0.0	.021
G	R04	(lysine)	0.0	—1.009
G	R05	(tryptophan)	—1.55	0.0
G	R06	(methionine)	0.0	—.732
G	R07	(cystine)	—.15	0.0
E	R08	(salt)	0.0	—.008
G	R09	(calcium)	—1.36	0.0
G	R10	(phosphorus)	0.0	—.218
G	R11	(niacin)	—4,191.9	0.0
G	R12	(riboflavin)	—1,200.07	0.0
G	R13	(pantothenic acid)	—365.02	0.0
G	R14	(vitamin B12)	0	—.075

   The shadow price for R01 indicates that reducing the equality restraint on the total weight by one pound would reduce the cost of the total mix by $.0177 per pound. However, the shadow price report assumes no change in the level of the other restraints.
5. A reduction of one pound in the protein restraint would not reduce the cost of the final mix.
6. The end product contains more protein, tryptophan, calcium, niacin, riboflavin, and pantothenic acid than specified when the original restraints were defined. Hence the slack activities in the basis were negative.
7. In the original model a maximum restraint was imposed on fiber content. Tightening the level of the restraint by one pound (permitting less fiber in the ration) would increase the cost of the mix by $0.021.

MODEL 12.4: *A Transportation Model*

Explanation

Like feed mix problems discussed in the preceding three models, the objective of transportation models is to meet a set of restraints at minimum cost. In the feed mix problem, ingredients must stay within certain maximum restraints (bulk and palatability) while equaling or exceeding the nutrient requirements specified, as minimum or equality restraints. The transportation model seeks to supply the product-deficit locations from surplus quantities available in other locations at minimum cost.

The example illustrates a structure appropriate for analyzing minimum cost patterns for transporting two varieties of seed corn among five locations with one-way and backhaul routes. Varieties are moved from surplus points to deficit locations. The beginning distribution of surpluses and deficits of each of the two varieties among the five locations is shown in Table 12.1.

Figure 12.1 shows the five locations and movements of surplus and deficit stocks among them. The five locations form a pentagon with warehouses G and O located in route from warehouse D to either warehouse S or L, thus allowing some backhaul routing. The cost of one-way hauls approximates 1.1 cents per mile per bushel; the cost of backhaul shipments is 0.7 cents per mile per bushel.

Restraints

R01 = a maximum restraint on surplus of variety 1 at warehouse S. The unit of restraint is bushels.
R02 = a maximum restraint on surplus of variety 1 at warehouse D. The unit of restraint is bushels.
R03 = an equality restraint on deficit of variety 1 at warehouse O. The unit of restraint is bushels.

TABLE 12.1: Distribution of Surplus and Deficit Quantities of Two Varieties of Seed Corn

Variety 1		Variety 2	
Location	Quantity (bu.)	Location	Quantity (bu.)
Surplus		Surplus	
S	3,000	G	4,000
D	10,000	L	3,500
Deficit		Deficit	
O	2,000	D	2,500
G	7,500	O	2,000
L	3,000	S	1,250

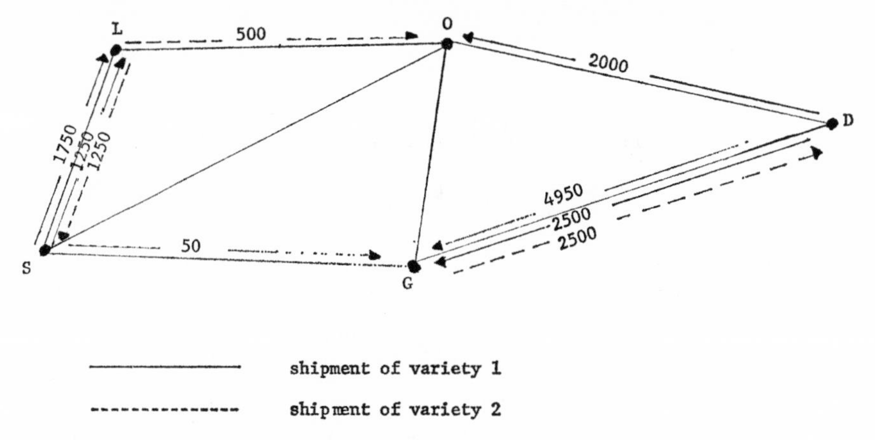

FIG. 12.1. Optimum shipment routes.

R04 = an equality restraint on deficit of variety 1 at warehouse G. The unit of restraint is bushels.
R05 = an equality restraint on deficit of variety 1 at warehouse L. The unit of restraint is bushels.
R06 = a maximum restraint on surplus of variety 2 at warehouse G. The unit of restraint is bushels.
R07 = a maximum restraint on surplus of variety 2 at warehouse L. The unit of restraint is bushels.
R08 = an equality restraint on deficit of variety 2 at warehouse D. The unit of restraint is bushels.
R09 = an equality restraint on deficit of variety 2 at warehouse O. The unit of restraint is bushels.
R10 = an equality restraint on deficit of variety 2 at warehouse S. The unit of restraint is bushels.

Activities

P01–P06 are shipping activities transporting one bushel of variety 1 between warehouses as follows:

P01 from S to O
P02 from S to G
P03 from S to L
P04 from D to O
P05 from D to G
P06 from D to L

P07–P12 are shipping activities transporting one bushel of variety 2 between warehouses as follows:

P07 from G to D
P08 from G to O
P09 from G to S
P10 from L to D

P11 from L to O
P12 from L to S

P13 = a shipping activity transporting one bushel of variety 1 from G to S and backhauling one bushel of variety 2 from S to G.
P14 = a shipping activity transporting one bushel of variety 1 from S to L and backhauling one bushel of variety 2 from L to S.
P15 = a shipping activity transporting one bushel of variety 1 from D to G and backhauling one bushel of variety 2 from G to D.
P16 = a shipping activity transporting one bushel of variety 1 from D to L and backhauling one bushel of variety 2 from L to D.
P17 = a shipping activity transporting one bushel of variety 1 from D to L and backhauling one bushel of variety 2 from L to D.
P18 = a shipping activity transporting one bushel of variety 1 from S to O and backhauling one bushel of variety 2 from G to S.

Points to Observe

1. The optimization process will yield the minimum cost pattern of transporting the two varieties among the five warehouses.
2. Upon optimization, the deficit quantities at all locations will have been eliminated and the least-cost mix of hauling activities for making the necessary transfers will have been defined.
3. All reasonable backhaul activities should be defined in the development of the mix of activities for the model. Likewise all possible one-way hauls should be specified as potential activities in the development of the transportation plan.
4. In defining row types, the deficit quantities are equality restraints. Thus in the optimization process all deficits will be eliminated. The surplus quantities at each location are treated as maximum restraints. Because the surplus exceeds the deficit for the complete system, a surplus will remain at one or more locations following optimizations. The slack activities for the appropriate restraints will appear in the basis signifying the location and quantity of "untransported" surplus.

Interpretation of Results

1. The total cost of transporting corn from the surplus to the deficit locations is $3,126.60.

MODEL 12.4

Row Type		B	P01	P02	P03	P04	P05	P06	P07	P08	P09	P10	P11	P12	P13	P14	P15	P16	P17	P18
N	C		.299	.191	.145	.210	.229	.393	.229	.108	.191	.393	.183	.145	.267	.203	.321	.550	.460	.375
L	R01	3,050	1	1	1										1	1				1
L	R02	10,000				1	1	1									1	1	1	
E	R03	2,000	1			1														1
E	R04	7,500		1			1								1		1			
E	R05	3,000			1			1								1		1	1	
L	R06	4,000							1	1	1				1		1			1
L	R07	3,500										1	1	1		1		1	1	
E	R08	2,500							1			1					1	1		
E	R09	2,000								1			1						1	
E	R10	1,250									1			1	1	1				1

TABLE 12.2: Output Report on Rows

Restraint	Level	Slack	Shadow Price
R01	3000		0.038
R02	9500	550	0.000
R03	2000		—0.210
R04	7500		—0.229
R05	3000		—0.183
R06	4000		0.075
R07	1750	1750	0.000
R08	2500		—0.167
R09	2000		—0.183
R10	1250		—0.058

2. After optimization, 550 bushels of variety 1 remain stored at warehouse D and 1,750 bushels of variety 2 remain at warehouse L. This excess supply results in the zero shadow prices for R02 and R07 (see Table 12.2).
3. The 0.038 shadow price on slack activity R01 indicates that the costs of meeting the deficit requirement on variety 2 could have been reduced by 3.8 cents had one more bushel been available at warehouse S. Alternatively, had one less bushel been available, the cost of meeting deficit requirements would have increased by 3.8 cents. The shadow price for R06 can be interpreted the same (if one less bushel of variety 2 were available, the cost would increase by 7.5 cents).
4. Increases in the quantities required at the deficit warehouses would result in a higher total shipping cost for the system as reflected by the shadow prices on restraints R03, R04, R05, R08, R09, and R10 (an increase of one bushel of variety 1 required at warehouse O would result in an added cost of .21 cents).
5. The hauling activities which entered the basis to form a minimum cost shipping pattern are as follows:

Activity	Level
P02	50
P03	1,750
P04	2,000
P05	4,950
P08	1,500
P11	500
P14	1,250
P15	2,500

6. The P14 activity represents shipment of 1,250 bushels of variety 1 from S to L and 1,250 bushels of variety 2 from L to S. The variety 1 deficit (3,000 bushels) at L is met by the backhaul shipment of P14 and the one-way haul of 1,750 bushels represented by activity

P03. (See the diagram of optimum shipping patterns in Figure 12.1.)

7. The activities which did not enter the basis and their shadow prices (cost penalties) are:

Activity	Shadow Price
P01	0.127
P06	0.210
P07	0.137
P09	0.208
P10	0.226
P12	0.087
P13	0.093
P16	0.200
P17	0.094
P18	0.220

8. The cost penalty associated with P01 indicates that shipment of one bushel of variety 1 from S to O would increase total shipping costs by $.127. Similar interpretation can be made for the other hauling activities which appear above.

MODEL 12.5: *A Transshipment Model*

Explanation

The transshipment model is similar in structure to the transportation model with the exception of the introduction of intermediate destinations from which the commodities are transported to final destinations (via shipping activities). The introduction of transfer rows permits a reduction in the number of activities while retaining all possible routes for shipment. In this model corn originates in two regions and is to be shipped to supply needs at twelve destinations. En route it passes through one of three intermediate ports. The objective is to define the mix of activities (shipment routes) that will minimize the cost of transporting corn from the producing regions to the destination points in the quantities required.

Activities

The procedure used in labeling the activities in this model departs from the convention followed in previous models. Typically the user has great flexibility in the name given the activities and restraints. The only limitation on labeling models processed with MPS/360 is that names cannot exceed eight alphabetic or numeric characters. Occasionally, when a model containing large numbers of activities and/or restraints is to be optimized, it is a convenience in interpreting output to employ names which enable the user to identify the nature of the activity upon sight. In this model the conventional P01, P02, . . . PX labels are not used but are replaced by several letters which refer to points of origin, transshipment, and destination. A similar naming system could be adopted to facilitate the identification of rows.

Each of the activities represents the shipment of one bushel of corn (C) from one point to another as indicated. Activities WMA–EMGL transport corn from the two producing regions to the three intermediate points. Activities AWSA–GLNSE ship the product from the three ports to the twelve destination points.

(WMA) from the Western Midwest center to the Atlantic
(WMG) from the Western Midwest center to the Gulf
(WMGL) from the Western Midwest center to the Great Lakes
(EMA) from the Eastern Midwest center to the Atlantic
(EMG) from the Eastern Midwest center to the Gulf
(EMGL) from the Eastern Midwest center to the Great Lakes
(AWSA) from the Atlantic to Western South America
(AESA) from the Atlantic to Eastern South America
(ANCE) from the Atlantic to North Central Europe
(AWA) from the Atlantic to Western Asia
(AEA) from the Atlantic to Eastern Asia
(ANAF) from the Atlantic to Northern Africa
(AWAF) from the Atlantic to Western Africa
(AEAF) from the Atlantic to Eastern Africa
(ASAF) from the Atlantic to Southern Africa
(ASSA) from the Atlantic to Southern South America
(ACAC) from the Atlantic to Central Atlantic Caribbeans
(ANSE) from the Atlantic to Northeastern and Southern Europe
(GWSA) from the Gulf to Western South America
(GESA) from the Gulf to Eastern South America
(GNCE) from the Gulf to North Central Europe
(GWA) from the Gulf to Western Asia
(GEA) from the Gulf to Eastern Asia
(GNAF) from the Gulf to Northern Africa
(GWAF) from the Gulf to Western Africa
(GEAF) from the Gulf to Eastern Africa
(GSAF) from the Gulf to Southern Africa
(GSSA) from the Gulf to Southern South America
(GCAC) from the Gulf to Central Atlantic Caribbeans
(GNSE) from the Gulf to Northeastern and Southern Europe

(GLWSA) from the Great Lakes to Western South America
(GLESA) from the Great Lakes to Eastern South America
(GLNCE) from the Great Lakes to North Central Europe
(GLWA) from the Great Lakes to Western Asia
(GLEA) from the Great Lakes to Eastern Asia
(GLNAF) from the Great Lakes to Northern Africa
(GLWAF) from the Great Lakes to Western Africa
(GLEAF) from the Great Lakes to Eastern Africa
(GLSAF) from the Great Lakes to Southern Africa
(GLSSA) from the Great Lakes to Southern South America
(GLCAC) from the Great Lakes to Central Atlantic Caribbeans
(GLNSE) from the Great Lakes to Northeastern and Southern Europe

Restraints

R01 is a transfer row for corn shipped to the Atlantic for subsequent shipment to final destinations.

R02 is a transfer row for corn shipped to the Gulf for subsequent shipment to final destinations.

R03 is a transfer row for corn shipped to the Great Lakes for subsequent shipment to final destinations.

Rows R04–R15 are equalities restraining the quantity demanded at each of the twelve destination points. The B column coefficient in each case is in terms of bushels.

R04 = Western South America
R05 = Eastern South America
R06 = North Central Europe
R07 = Western Asia
R08 = Eastern Asia
R09 = Northern Africa
R10 = Western Africa
R11 = Eastern Africa
R12 = Southern Africa
R13 = Southern South America
R14 = Central Atlantic Caribbeans
R15 = Northeastern and Southern Europe

The last two restraints limit the supply of corn that can originate in the two regions of production, Western Midwest states and Eastern Midwest states. The unit of each restraint is bushels.

R16 = a restraint on the supply (production) of corn in the Western Midwest states.
R17 = a restraint on the supply (production) of corn in the Eastern Midwest states.

MODEL 12.5

	Row Type	B	WMA	WMG	WMGL	EMA	EMG	EMGL	AWSA	AESA	ANCE	AWA	AEA	ANAF	AWAF	AEAF	ASAF	ASSA	ACAC	ANSE	GWSA	GESA	GNCE	GWA	GEA	GNAF	GWAF	GEAF	GSAF	GSSA	GCAC	GNSE	GLWSA	GLESA	GLNCE	GLWA	GLEA	GLNAF	GLWAF	GLEAF	GLSAF	GLSSA	GLCAC	GLNSE
C	N		.37	.28	.16	.26	.28	.09	.40	.49	.17	.52	.39	.54	.59	.73	.55	.76	.40	.26	.45	.69	.32	.80	.23	.78	.80	.85	.75	.81	.59	.39	.53	.58	.20	.59	.47	.56	.67	.78	.64	.83	.54	.29
R01	L	0	-1			-1			1	1	1	1	1	1	1	1	1	1	1	1																								
R02	L	0		-1			-1														1	1	1	1	1	1	1	1	1	1	1	1												
R03	L	0			-1			-1																									1	1	1	1	1	1	1	1	1	1	1	1
R04	E	1,949,968							1												1												1											
R05	E	1,214,008								1												1												1										
R06	E	242,383,920									1												1												1									
R07	E	10,359,226										1												1												1								
R08	E	11,004,182											1												1												1							
R09	E	666,098												1												1												1						
R10	E	3,523,195													1												1												1					
R11	E	4,707,036														1												1												1				
R12	E	5,936,537															1												1												1			
R13	E	78,114,567																1												1												1		
R14	E	3,260,609																	1												1												1	
R15	E	134,785,794																		1												1												1
R16	L	249,214,977	1	1	1																																							
R17	L	304,596,100				1	1	1																																				

Points to Observe

1. Two points of origin are designated in this model, a point within the Western Midwest and a point within the Eastern Midwest. There are twelve destination points. Specifying additional points would make the model more nearly approach reality, particularly in respect to the location of production. However, the number of activities increases rapidly with increases in either points of origin or destination.
2. As in the previous model, the quantities demanded at each of the final destinations have been treated as equality restraints.
3. The use of transfer rows follows the pattern established in previous models. The coefficients at the intersection of each of the intermediate transport activities (WMA–EMGL) and the rows designating the area of origin (R16 and R17) are plus ones. The coefficients at the intersection of the intermediate transport activities and the rows representing the intermediate (transshipment) points (R01–R03) are negative ones.

PROCESSING MINIMIZATION MODELS

The control cards needed to process models designed to minimize some objective function are listed on page 200. Only one card differs from the deck used in maximization. In the latter case the letters MAX appear on the SETUP card. For the minimization process MAX is deleted and only the letters SETUP appear.

Provisions for artificial activities, where appropriate, need to be built into the model. Because minimization problems frequently involve a mix of maximum, minimum, and equality restraints, special care should be exercised to see that restraints are properly labeled as to type.

SIMPLEX SOLUTION OF RATION PROBLEM

Before attempting to trace the steps to the solution of a minimization problem, the reader should review the application of the simplex method to the crop production problem in Chapter 2. Although the latter involves a maximization model, the routine followed in the minimization procedure has many similarities. The solution to the ration problem presented in Model 12.1 has been developed step by step. Table 12.3 and the explanation of procedure relate to this problem.

MINIMIZATION

Job Control Language and Control Program Cards

```
JOB CONTROL LANGUAGE CARDS


               PROGRAM
               INITIALZ
               MOVE(XDATA,'ECON430')
               MOVE(XPBNAME,'PBFILE')
               MVADR(XMAJERR,UNB)
               MVADR(XDONFS,NOF)
               CONVERT
               SETUP
               MOVE(XRHS,'B')
               MOVE(XOBJ,'C')
               PRIMAL
               SOLUTION
               EXIT
NOF            TRACE
UNB            EXIT
               PEND
/*


JOB CONTROL LANGUAGE CARDS


NAME                ECON430


ENDATA
/*
```

1. In this example the original matrix is formed with the artificial activities in the basis. By contrast, in the maximization problem all restraints were maximum restraints, and the disposal activities appeared in the original and feasible but nonoptimal solution. Thus the initial solution is as follows:

   A01 = 4,500.00
   A02 = 27,000.00
   A03 = 54,000.00

   It should be noted that all activities appearing in the initial solution are artificial because all restraints were minimum restraints and hence coefficients for all disposal activities were negative. The disposal activities for maximum restraints in the minimization model appear in the original solution in the same manner as in maximization models.

2. After the original matrix has been formed, the first step is to select an outgoing column. The selection criterion is the *highest positive* Z-C value.
3. The Z values for the original matrix can be computed by multiplying the coefficients in each column by the price of the activity appearing in the corresponding row in the B column and summing the products. For the sake of simplicity, we have assigned a price of $10,000 to each artificial activity. Thus the Z value for P01 is determined as follows:

	Activity Price	P01 Coefficients	Product
A01	10,000	.130	1,300
A02	10,000	.480	4,800
A03	10,000	2.160	21,600
			$27,700

and Z-C = 27,700 — .010 = 27699.99

By a similar process the Z-C coefficient can be computed for each column. Inspection of the first section of the solution indicates that P01 has the highest positive Z-C value and hence is selected as the outgoing column.

4. The row with the smallest R column coefficient is the outgoing row. The ratio column is formed as in the maximization procedure by dividing the B column elements by the corresponding coefficients in the outgoing column. Thus:

A01 4,500 ÷ .130 = 34,615.385
A02 29,000 ÷ .480 = 56,250.000
A03 54,000 ÷ 2.160 = 25,000.000

and A03 becomes the outgoing row.

5. The procedure for forming the elements in Section II is exactly the same in this illustration as in the maximization process. The incoming row is determined as in Chapter 2 by dividing all elements by the pivot. Other rows can be computed from the relationship O — (I × P) = N described in Chapter 2 where O, I, P, and N are defined exactly as in the maximization example.
6. The iterative process continues, selecting the outgoing column on the basis of the most positive Z-C coefficient and the outgoing row on the basis of the smallest R value. The process is terminated (i.e., the model has been optimized) when all Z-C values become negative or zero.

7. The reader should note that the process extends to five iterations and six sections before an optimum solution is achieved. This is characteristic of simplex solutions where the model contains several minimum restraints and their accompanying artificial activities. In this illustration three iterations are required before we achieve a basis in Section IV void of artificial activities.
8. The value of the objective function (value of the program) is high in the first solution and diminishes until we reach the minimum cost ration in Section VI.

ALGEBRAIC FORMULATION OF MINIMIZATION REQUIREMENTS

The algebraic presentation which follows is based on the ration problem of Model 12.1.

Let:

X_A = units of alfalfa hay
X_C = units of corn
X_S = units of soybean oil meal

Then, from Model 12.1

$$.13X_A + .016X_C + .07X_S \geq 4{,}500 \text{ units of protein} \quad (12.1)$$
$$.48X_A + .73X_C + .45X_S \geq 27{,}000 \text{ units of TDN} \quad (12.2)$$
$$2.16X_A + 0X_C + 0X_S \geq 54{,}000 \text{ units of vitamin A} \quad (12.3)$$

Furthermore:

$$X_A \geq 0 \quad (12.4)$$
$$X_C \geq 0 \quad (12.5)$$
$$X_S \geq 0 \quad (12.6)$$

The above conditions reflect the impossibility of feeding a negative quantity of some ration component.

The next step is to form equations from inequalities (12.1), (12.2), and (12.3) above. Because all three are "equal to or greater than" restraints, the slack activities provide for overfulfillment of the restraint and carry a negative sign. Hence, in the example above more protein can be supplied from the three sources than the minimum 4,500 units required. The magnitude of any overfulfillment would be reflected in the slack variable:

$$.13X_A + .074X_C + .45X_S - X_P = 4{,}500 \quad (12.7)$$
$$.48X_A + .73X_C + .78X_S - X_T = 27{,}000 \quad (12.8)$$
$$2.16X_A + 0X_C + 0X_S - X_V = 54{,}000 \quad (12.9)$$

BLE 12.3: Simplex Solution to Beef Cow Ration Problem

Section I

B	P01	P02	P03	R01	R02	R03	A01	A02	A03	
4,500.000	.130	.074	.450	-1.000	0.000	0.000	1.000	0.000	0.000	G
27,000.000	.480	.730	.780	0.000	-1.000	0.000	0.000	1.000	0.000	G
54,000.000	2.160	0.000	0.000	0.000	0.000	-1.000	0.000	0.000	1.000	G
	.010	.016	.070	0.000	0.000	0.000	10,000.000	10,000.000	10,000.000	N
85,500,000.	27,699.99	8,039.984	12,299.970	-10,000.000	-10,000.000	-10,000.000	0.000	0.000	0.000	

UE OF OBJECTIVE FUNCTION = $855,000,000.00

Section II

B	P01	P02	P03	R01	R02	R03	A01	A02	A03
1,250.000	0.000	.074	.450	-1.000	0.000	.060	1.000	0.000	-.062
15,000.000	0.000	.730	.780	0.000	-1.000	.222	0.000	1.000	-.222
25,000.000	1.000	0.000	0.000	0.000	0.000	-.463	0.000	0.000	.463
162,500,250.	0.000	8,040.000	12,300.000	-10,000.000	-10,000.000	2,824.100	0.000	0.000	-12,824.000

UE OF OBJECTIVE FUNCTION = $162,500,250.00

Section III

B	P01	P02	P03	R01	R02	R03	A01	A02	A03
2,777.800	0.000	.164	1.000	-2.222	0.000	.134	2.222	0.000	-.134
12,833.000	0.000	.602	0.000	1.733	-1.000	.118	-1.733	1.000	-.118
25,000.000	1.000	0.000	0.000	0.000	0.000	-.463	0.000	0.000	.463
125,330,444.45	0.000	6,017.300	0.000	17,333.000	-10,000.000	1,179.000	-27,333.000	0.000	-11,179.000

UE OF OBJECTIVE FUNCTION = $128,330,444.45

Section IV

B	P01	P02	P03	R01	R02	R03	A01	A02	A03
19,231.000	0.000	.936	1.000	0.000	-1.282	.285	0.000	1.282	-.285
7,403.800	0.000	.347	0.000	1.000	-.577	.068	-1.000	.577	-.068
25,000.000	1.000	0.000	0.000	0.000	0.000	-.463	0.000	0.000	.463
1,596.17	0.000	.050	0.000	0.000	-.090	.015	-10,000.000	-9,999.900	-10,000.000

UE OF OBJECTIVE FUNCTION = $1,596.17

Section V

B	P01	P02	P03	R01	R02	R03	A01	A02	A03
20,548.000	0.000	1.000	1.069	0.000	-1.370	.304	0.000	1.370	-.304
270.550	0.000	0.000	-.371	1.000	-.101	-.038	-1.000	.101	.038
25,000.000	1.000	0.000	0.000	0.000	0.000	-.463	0.000	0.000	.463
578.77	0.000	0.000	-.053	0.000	-.022	.003	-10,000.000	-10,000.000	-10,000.000

UE OF OBJECTIVE FUNCTION = $578.77

Section VI

B	P01	P02	P03	R01	R02	R03	A01	A02	A03
67,500.000	0.000	3.285	3.510	0.000	-4.500	1.000	0.000	.045	-1.000
2,812.500	0.000	.124	-.239	1.000	-.271	0.000	-1.000	.271	0.000
56,250.000	-1.000	1.521	1.625	0.000	-2.083	0.000	0.000	2.083	0.000
562.50	0.000	-.001	-.054	0.000	-.021	0.000	-10,000.000	-10,000.000	-10,000.000

UE OF OBJECTIVE FUNCTION = $562.50

where X_P, X_T, and X_V are slack variables for protein, TDN, and vitamin A restraints respectively.

If we were to begin by assigning all real activities a zero value as we did in the maximization process, then

$$X_P = -4{,}500$$
$$X_T = -27{,}000$$
$$X_V = -54{,}000$$

These results are incompatible with conditions specified in (12.4), (12.5), and (12.6) above. To circumvent this difficulty we add an artificial variable to each equation. We now have:

$$.13X_A + .074X_C + .45X_S - X_P + A_P = 4{,}500 \quad (12.10)$$
$$.48X_A + .73X_C + .78X_S - X_T + A_T = 27{,}000 \quad (12.11)$$
$$2.16X_A + 0X_C + 0X_S - X_V + A_V = 54{,}000 \quad (12.12)$$

where A_P, A_T, and A_V are artificial activities.

We assign zero values to all variables except the three artificial variables. This manipulation provides the original solution:

$$A_P = 4{,}500$$
$$A_T = 27{,}000$$
$$A_V = 54{,}000$$

The solution is a wholly artificial one and no attempt should be made to give it an economic interpretation, since it is nothing more than a mechanism by which the simplex method can be initiated.

We seek to minimize the objective function:

$$C = .01X_A + .016X_C + .07X_S + MA_P + MA_T + MA_V \quad (12.13)$$

where C = cost of providing winter feed for 25 beef cows for 120 days.

X_A, X_C, and X_S are as defined previously, and
M = a very large positive number.

CHAPTER THIRTEEN

OTHER MAXIMIZATION APPLICATIONS

The four sections of this chapter are presented to illustrate applications outside the range of whole farm planning as presented heretofore. The first section with its model illustrates a method of handling the time dimensions of supply and demand for forage in whole farm planning. The second section illustrates a method of determining the optimum level of water application when several cropping alternatives are presented. The third section illustrates two methods or models for blending corn to meet minimum and maximum standards specified for a grade. The last section illustrates a model for determining the optimum level of livestock production and the optimum mix of feed ingredients to meet the nutritional needs of the livestock produced.

MODEL 13.1: *Forage Planning Model*

Planning the forage program as approached below involves building a whole farm planning model which gives particular attention to forage producing and forage using activities. This problem is given special treatment because of the critical role played by the time dimension in a realistic analysis of optimum forage production. The model assumes that beef raising is the only feasible livestock activity. The objective function is maximization of income. Achieving this objective implies that the forage program will be organized to provide forage for the beef cow by the most efficient means, given the restraints and the range of forage alternatives available.

To treat the forage problem adequately the model must provide for: (1) choice among a wide range of forage crops, (2) choice among levels of fertilizer use, (3) activities which permit alternative grazing including delayed grazing, (4) a range of harvesting and storage alternatives, and (5) a breakdown on a monthly basis of quantities forthcoming from forage supplying activities and of quantities required by forage using activities. This model illustrates all of these features.

Activities

P01 = a cow-calf production activity. The unit of activity is one cow and .9 calf (90% calf crop) with the calf sold at 475 lb.

P02 = a labor hiring activity. The unit is one hour.

P03 = a nitrogen purchasing activity. The unit of activity is one pound.

P04 = a phosphorus purchasing activity. The activity unit is one pound.

P05 = a potash purchasing activity. The unit of activity is one pound.

P06–P07 = bird's-foot trefoil pasture supplying activities on class II and III land respectively. The unit of each activity is one acre being continuously grazed.

P08–P09 = bird's-foot trefoil pasture supplying activities on class II and III land respectively. The unit of each activity is one acre being stockpiled for late grazing.

P10–P11 = bird's-foot trefoil pasture supplying activities on class II and III land respectively. The unit of each activity is one acre being grazed early, then stockpiled until fall.

P12–P13 = bird's-foot trefoil forage supplying activities. The unit of each activity is one acre with the first cutting being harvested and the remaining forage being pastured.

P14–P25 = harvested forage supplying activities. The unit of each activity is one cow-day with TDN requirements as follows:

Month	Lb. TDN	Month	Lb. TDN
January	8	July	19
February	8	August	16
March	8	September	16
April	8	October	16
May	19	November	14
June	19	December	8

P26–P27 = bird's-foot trefoil renovation activities on class II and III land respectively. The unit of each activity is one acre.

P28 = an alfalfa-grass pasture supplying (continuous grazing) activity. The unit of activity is one acre.

P29 = an alfalfa-grass forage supplying activity. The unit of activity is one acre with the first two cuttings harvested.

P30 = an alfalfa-grass renovation activity. The unit is one acre.

P31 = a smooth bromegrass forage supplying activity. The unit of activity is one acre of smooth bromegrass with 240 lb. nitrogen applied annually. The first cutting is harvested.

P32 = a smooth bromegrass supplying activity which provides for continuous grazing. The unit of activity is one acre with 120 lb. nitrogen applied annually.

P33 = an activity renovating smooth bromegrass. The unit of activity is one acre.

P34 = a corn silage production and harvesting activity. The unit is one acre.

Restraints

C = the objective function.
R01 = a restraint on class I land. Each B column unit is one acre.
R02 = a restraint on class II land. Each B column unit is one acre.
R03 = a restraint on class III land. Each B column unit is one acre.
R04–R15 = transfer rows for feed being transferred to the cow-calf activity for each month. The unit of transfer is cow-calf days.
R16 = a restraint on labor. The unit of restraint is hours.
R17 = a transfer row for nitrogen. The unit of transfer is pounds.
R18 = a transfer row for phosphorus (P_2O_5). The unit of transfer is pounds.
R19 = a transfer row for potash (K_2O). The unit of transfer is pounds.
R20 = a transfer row for renovated bird's-foot trefoil produced on class II land. The transfer unit is acres.
R21 = a transfer row for renovated bird's-foot trefoil produced on class III land. The unit of transfer is acres.
R22 = a harvested TDN transfer row. The unit of transfer is pounds.
R23 = a transfer row for renovated alfalfa produced on class II land. The unit of transfer is acres.
R24 = a transfer row for renovated smooth bromegrass produced on class II land. The unit of transfer is acres.

Points to Observe

1. The complete range of reasonable forage crop alternatives should be defined for each land class. In this illustration a full range has not been presented. However, alternatives have been illustrated with bird's-foot trefoil being seeded on class II and III land. Several grazing, harvesting, and fertilizer application alternatives are also presented.
2. The mix of land classes is reflected in the B column coefficients for R01, R02, and R03 (land restraints). In the planning situation being dealt with here there are 40 acres of class I land, 350 acres of class II, and 250 acres of class III. Activities P26, P27, P30, and P33 provide for renovation of the bird's-foot trefoil, alfalfa, and smooth bromegrass forage crops. These activities must enter into the basis before the grazing and/or harvesting activities can enter at nonzero levels. This condition is insured by the maximum constraints (R20, R21, R23, and R24). For example, the entry of activity P26 provides eight units of R20 to be used by P06, P08, P10, or P12.

	Row type	B	P01	P02	P03	P04	P05	P06	P07	P08	P09	P10	P11	P12	P13	P14	P1
C	N		125	-2.25	-.09	-.09	-.04	-1.25	-1.25	-1.25	-1.25	-1.25	-1.25	-11.8	-11.03		
R01	L	40															
R02	L	350						1		1		1		1			
R03	L	250							1		1		1		1		
R04	L		35.75													-1	
R05	L		32.2														-
R06	L		35.75														
R07	L		34.5									-8.8	-8				
R08	L		35.75					-5.5	-5.			-38.5	-35				
R09	L		34.5					-39.6	-36	-20.9	-19						
R10	L		35.75					-36.3	-33	-49.5	-45	-19.8	-18	-17.6	-16.		
R11	L		35.75					-33	-30	-35.2	-32	-35.2	-32	-33	-30		
R12	L		34.5					-20.9	-19	-20.9	-19	-27.5	-25	-27.5	-25		
R13	L		35.75					-11	-10	-14.3	-13	-13.2	-12	-13.2	-12		
R14	L		34.5														
R15	L		35.75														
R16	L			-1				.3	.3	.3	.3	.3	.3	1.98	1.83		
R17	L				-1												
R18	L					-1		30	30	30	30	30	30	30	30		
R19	L						-1	30	30	30	30	30	30	30	30		
R20	L							1		1		1		1			
R21	L								1		1		1		1		
R22	L													-1042.7	-952	8	8
R23	L																
R24	L																

3. The length of time the renovated grasses and legumes will remain in production before renovation or reestablishment is undertaken must be known. In this model it was assumed that bird's-foot trefoil would remain in production eight years. Therefore, the coefficient of P26 in R20 equals —8 and the coefficient of P27 in R21 equals —8. It was estimated that alfalfa would remain productive for three years and bromegrass for ten before reestablishment became necessary.
4. The TDN requirements must reflect accurately the nutrient requirements of the cow and calf throughout the year. The P01 activity assumes a spring calving schedule; thus a high TDN intake per cow-day is assumed from May through November when the calf is nursing (see activities P14–P25 for TDN requirements per cow-day for each month). It is also assumed that the cow can

P16	P17	P18	P19	P20	P21	P22	P23	P24	P25	P26	P27	P28	P29	P30	P31	P32	P33	P34
										-53.46	-53.46	-2.70	-19.07	-48.04	-18.54	-1.85	-53.11	-74.80
																		1
										1		1	1	1	1	1	1	
											1							
-1																		
	-1															-6		
		-1										-25				-32		
			-1									-32				-32		
				-1						-10	-10	-25		-10		-18	-10	
					-1					-15	-15	-25	-28	-15		-15	-15	
						-1						-10	-10		-16	-20		
							-1			-10	-10	-23	-27	-10	-24	-10	-10	
								-1		-10	-10			-10	-39		-10	
									-1						-38			
										3.03	3.03	.6	3.06	3.03	3.04	.3	3.03	8.45
															240	120		150
												40	40		40	40		40
												60	60		40	40		40
										-8								
											-8							
8	8	19	19	19	16	16	16	14	8	-664	-664		-1624	-664	-1808		-664	-6656
												1	1	-3				
															1	1	-10	

maintain her weight on eight pounds of TDN per day during the period from December to April. TDN requirement during this period is set at a minimum to minimize the amount of harvested forages being consumed. Alternative assumptions in respect to the distribution of TDN requirements would result in a different mix of activities.

5. Forage which is harvested in the model is placed in the transfer row (R22) and is assumed to be of homogeneous quality and available for feeding throughout the year. Activities P14–P25 supply each of the twelve months with TDN requirements from R22. A transfer row or rows separate from R22 must be provided to transfer forage or refuse that is available only during particular months. Using material of this type also requires that appropriate feeding transfer activities be provided.

Interpretation of Results

1. The optimum plan yields a value of the program of $20,360.59 with the cow-calf activity in the plan at a level of 326.27 cow units.
2. The average number of acres of perennial grasses and legumes being reestablished each year is 15.05 acres of bird's-foot trefoil on class II land, 27.78 acres of bird's-foot trefoil on class III land, and 19.50 acres of smooth bromegrass.
3. The acres of forage produced are as follows:

Forage Crop	Land Class	Harvesting and/or Grazing Practices	Acres
Bird's-foot trefoil	II	continuous graze	80.19
Bird's-foot trefoil	III	continuous graze	222.22
Bird's-foot trefoil	II	graze early, then stockpile	40.24
Smooth bromegrass	II	harvest 1 cutting, then graze	195.02
Corn silage	I	harvest	40.00

4. The number of cow-days of harvested feed being fed and the value of the feed for one cow unit for one day are as follows:

Month	Quantity of Harvested Feed Fed	Value
	(cow-days)	($ per cow-day)
January	11,664.1	.217
February	10,505.9	.217
March	11,664.1	.217
April	10,902.2	.217
May	8,562.9	.515
June	80.9	.515
July	0.0	.149
August	0.0	.064
September	1,131.3	.433
October	2,724.9	.433
November	3,027.1	.379
December	4,253.3	.217

5. The value of the product marketed through the cow herd from the last acre for each of the land classes is $67.76 for class I, $30.92 for class II, and $27.32 for class III.
6. The income penalties on activities not in the plan are as follows:

Activity	Income Penalty
P08	$ 8.91
P09	8.10
P12	8.28
P13	7.61
P28	11.07
P29	20.13
P32	4.78

MODEL 13.2: *Irrigation Applications*

Models may be constructed to aid in decisions involving the allocation of water. Where irrigation systems are organized around projects or districts, some of the most important allocation decisions are made by a central authority. The amount of water to be made available by time periods during the growing season is an example of such a decision. Sometimes, particularly in underdeveloped countries, the decisions made by the project administrators may encompass the amount of land to be brought under irrigation and the mix of crops to be produced. Linear programming is especially useful in clarifying the decisions required of the central authority, but to provide meaningful guidelines such models must be large and complex.

Under most circumstances in the United States decisions regarding water use are made at the farm level with the exception of the seasonal pattern of water release, which would in some districts be regulated at the project level. The models in this chapter focus on the individual farm, although many of the features discussed are also applicable to project models.

Some Characteristics of Irrigation Models

To develop realistic models encompassing water allocation, the planner must divest himself of the misconception that each crop requires a single level of water use. On the contrary, every crop can be produced within a wide range of water levels. When moisture is not supplemented by rainfall, low applications of water per unit of land may result in no output. At some point, however, as more water is applied, a small output will be forthcoming. Next follows a range within which output per unit of water increases. Then as water use increases still more, each added unit of water results in a lower marginal increment of output. It would not make economic sense to consider levels of water use below the point where the average yield per unit of water is at a maximum. If restraints on water were so severe that applications sufficient to reach the maximum water yield were not possible on all units of land available, it would be more profitable to reduce the land area under irrigation. Water use can also be pushed to the point where its yield at the margin is zero. Finding the optimum level between the point of maximum average water yield and zero marginal yield is the heart of the water allocation problem. It becomes highly complex where several crops are reasonable alternatives, because each crop responds uniquely to water use.

The planner attempting to construct a model treating water allocation will be confronted immediately with the scarcity of information now in existence concerning the response of different crops to water. Although this problem is not peculiar to water coefficients, relative to the importance and the complexity of the water allocation problem, there is no other decision area in agriculture with such a void of information. The single-level approach to water use makes economic sense only where water is a free good and hence should be used at the point where its marginal yield is zero and per acre yields are maximized. Few decision situations are of this nature. Therefore, in the search for realistic models the planner must make do with the best estimates of water response he can obtain, however gross and unsatisfactory they may be, knowing that answers obtained in this way are more likely to be useful than those based on a single-level assumption.

Explanation

The model provides for growing three different crops. Each crop may be produced under four levels of water use (including zero level). The activities presented are highly aggregated in that they include raising, harvesting, selling, and irrigating. Usually in constructing crop production models it is desirable to define each of these functions as separate activities. However, several functions have been combined in this model to conserve space.

Activities

P01 = an alfalfa growing and harvesting activity with a high level of irrigation. The unit of activity is one acre.
P02 = an alfalfa growing and harvesting activity with a medium level of irrigation. The unit of activity is one acre.
P03 = an alfalfa growing and harvesting activity with a low level of irrigation. The unit of activity is one acre.
P04 = an alfalfa growing and harvesting activity with no irrigation. The unit of activity is one acre.
P05 = a corn growing and harvesting activity with a high level of irrigation. The unit of activity is one acre.
P06 = same as P05 except irrigation is at medium level.
P07 = same as P05 except irrigation is at low level.

P08 = same as P05 except there is no irrigation.
P09 = a wheat growing and harvesting activity with a high level of irrigation. The activity unit is one acre.
P10 = same as P09 except irrigation is at medium level.
P11 = same as P09 except irrigation is at low level.
P12 = same as P09 except there is no irrigation.
P13 = a hay selling activity. The unit of activity is one ton.
P14 = a corn selling activity. The unit of activity is one bushel.
P15 = a wheat selling activity. The unit of activity is one bushel.
P16 = a capital borrowing activity. The unit of activity is one dollar.
P17 = a labor hiring activity for months of March, April, May. The unit of activity is one hour.
P18 = a labor hiring activity for months of June, July, August. The unit of activity is one hour.
P19 = a labor hiring activity for months of September, October, November. The unit of activity is one hour.

Restraints

R01 = a land restraint. Each B column unit is one acre.
R02 = a labor restraint for December, January, and February. Each B column unit is one hour.
R03 = a labor restraint for March, April, May. Each B column unit is one hour.
R04 = a labor restraint for June, July, August. Each B column unit is one hour.
R05 = a labor restraint for September, October, November. Each B column unit is one hour.
R06 = an April irrigation water restraint. Each B column unit represents one acre-inch of water.
R07 = same as R06 except a May water restraint.
R08 = same as R06 except a June water restraint.
R09 = same as R06 except a July water restraint.
R10 = same as R06 except an August water restraint.
R11 = same as R06 except a September water restraint.
R12 = a hay transfer row. Each B column unit is one ton.
R13 = a corn grain transfer row. The transfer unit is one bushel.
R14 = a wheat grain transfer row. The transfer unit is one acre.
R15 = a capital transfer row. The transfer unit is one acre.
R16 = a restraint on total amount of water available for the growing season. The B column quantity includes expected rainfall as well as irrigation water. Each B column unit represents one acre-inch.

Points to Observe

1. Four levels of irrigation, one of which is zero, are defined by the coefficients in rows R06–R11. Any number of levels could be defined, but the planner will find specifying coefficients for four levels a sufficient challenge in view of the paucity of research results in water use. The pattern in which the water is distributed through the growing season is undoubtedly of great importance,

MODEL 13.2

	Row Type	B Column	P01	P02	P03	P04	P05	P06	P07	P08	P09	P10	P11	P12	P13	P14	P15	P16	P23	P24	P25
C	N		-29.91	-22.82	-16.98	-12.65	-46.38	-41.20	-36.09	-35.02	-14.98	-13.18	-12.66	-10.98	20.50	1.15	1.35	+.08	-2.35	-2.35	-2.35
R01	L	480	1	1	1	1	1	1	1	1	1	1	1	1							
R02	L	600																			
R03	L	750	1.82	1.97	1.47	1.07	3.02	3.00	2.96	1.27	3.75	3.75	3.70	.8					-1		
R04	L	700	7.84	7.09	6.34	4.89	3.36	2.91	2.90	1.41	3.65	3.65	3.50	1.6						-1	
R05	L	750	1.63	1.43	.48	.28	2.96	2.96	2.46	1.66											-1
R06	L	1,600	4.0	4.0	3.5		2.86	2.83	2.80		2.29	2.60	2.5	.01							
R07	L	1,800	4.0	4.0	3.5	.04				.99	4.58	3.90	2.5	.93							
R08	L	1,800	4.0	4.0	3.5	2.62	2.86	2.83	2.8	2.46	4.58	3.90	2.5	2.77							
R09	L	2,900	8.0	4.0		4.66	5.72	5.66	5.6	5.94	4.58	2.60	2.5	4.88							
R10	L	2,600	4.0			2.57	5.72	2.83	2.8	3.14				.69							
R11	L	1,500	4.0	4.0	3.5		2.86	2.83													
R12	L		-4.2	-3.4	-2.4	-1.6									1						
R13	L						-108.7	-98.2	-75.5	-58.7						1					
R14	L										-41	-36	-31	-23			1				
R15	L		19.14	19.14	19.14		19.14	19.14	19.14		19.14	19.14	19.14					-1			
R16	L	9,400	28.00	20.00	14.00	9.89	20.02	16.98	14.0	12.53	16.03	13.00	10.00	9.28							

but here again there is little research to provide guidelines for forming coefficients.

2. For each crop production activity, both monthly and yearly water requirements were defined. The yearly requirement is the sum of the monthly requirements. As the amount of water applied decreased from a high to a zero level, it was assumed that one or more water applications were eliminated which resulted in decreases in both monthly and yearly applications of water to the crop. In those cases where the monthly water coefficient was zero, the plants used water either from previous irrigation or from rainfall. To cite an example: P05, a corn growing and harvesting activity with a high level of irrigation, received 2.86 acre-inches of water from which sufficient moisture was retained in the soil to support crop growth in May following planting.
3. The pattern of B column coefficients by time periods depends on the water source. This model is based on well irrigation. There is a limit to the amount of water that can be pumped during any one time period—in this case any one month. Restraints R06–R11 represent the monthly limits. In addition, the total amount which can be drawn from the well during the course of a year is limited. R12 is the restraint on total withdrawals. Note that this amount is considerably less than the sum of the monthly restraints.
4. Where water is obtained from a reservoir, the formation of the appropriate coefficients in the B column of the water restraints can be considerably more complicated. The features necessary to model the reservoir as a source of water are essentially the same for an individual operator or a central authority administering water resources for an irrigation district. The entry for the B column coefficient for the beginning of the period (R06 in this example) would be the total amount expected in the reservoir at the beginning of the time period. Water would move from one time period to another through a set of water transfer activities where loss through evaporation and seepage would be reflected in the transfer coefficients. Thus, an acre-inch in m_1 might decrease to .85 acre-inches when transferred to m_2. Similarly, existing supplies in the reservoir may be supplemented during the season through rainfall and its resulting runoff. This requires a set of reservoir supplementing activities either restrained by an equality or bounded. Finally, prudence may lead the planner to set some minimum greater than zero on the quantity of water he would expect to leave in the reservoir at the end of the production

period. Where this is the case a carry-over activity should be formed which is restrained to the desired minimum.

MODELS 13.3 AND 13.4: *Grain Blending Applications*

Description of the Problem

Unlike the minimization problems presented in Chapter 12, this model has an objective function of maximization of income. The problem involves blending several bins of corn, each failing to meet one or more of the specifications for No. 2 grade, to obtain a blend that meets minimum weight requirements while staying within the maximum restraints on moisture, foreign matter, and damage. The standards for grading are based on the Winchester bushel; the purchase and sale of grain are based on a legal bushel defined as 56 pounds. The Winchester bushel, a bushel by volume, must weigh 54 pounds to meet minimum "test weight" requirements for No. 2 grade.

Other specifications for No. 2 corn are: (a) moisture content may not exceed 15.5%; (b) damaged material may not exceed 5%; (c) heat-damaged material may not exceed .2%; and (d) foreign material may not exceed 3%. When any of these specifications is not met, the purchase price based on the legal bushel is reduced by discounts established in the market. The discounts commonly applied are:

1. Moisture—two cents per bushel for each percent moisture in excess of 15.5%.
2. Test weight—one cent per bushel for each pound under 54 per Winchester bushel.
3. Damage—0.5 cent per percent damage in excess of 5%.
4. Foreign material—one cent for each additional percent or fraction thereof in excess of 3% up to 5%; then 2 cents for each additional percent or fraction thereof above 5%.

Grading specifications, price per bushel, and number of bushels for each of the eleven bins available for blending are presented in Table 13.1.

Activities

P01–P11 = activities which supply grain for blending from bins 1–11 respectively. The unit of each activity is one Winchester bushel.

P12 = a screening activity. The unit is removing (screening) 1% foreign material from one Winchester bushel.

P13 = a drying activity. The unit is removing 5% moisture from one Winchester bushel.

TABLE 13.1: Grade Specifications, Prices, and Number of Bushels Available for Blending

Bin	Test Weight	Moisture	Damage	Foreign Material	Price per Legal Bushel	Number of Legal Bushels	Number of Winchester Bushels
	(lb.)	(%)	(%)	(%)			
1	54.5	15.7	3.0	2.6	1.03	17,517.87	18,000
2	55.0	15.0	1.4	2.9	1.04	17,678.57	18,000
3	25.0	13.5	4.2	4.3	1.00	16,714.29	18,000
4	55.0	14.7	2.6	2.9	1.04	17,678.57	18,000
5	55.9	14.5	14.9	5.4	.95	17,967.86	18,000
6	53.9	16.0	3.0	2.8	1.02	17,196.43	18,000
7	54.5	15.7	3.0	2.7	1.03	17,517.87	18,000
8	52.0	15.0	3.8	4.0	1.01	16,714.29	18,000
9	53.0	15.3	2.0	1.7	1.03	17,035.71	18,000
10	53.0	14.3	7.9	9.6	.895	17,035.71	18,000
11	53.0	13.5	8.9	3.0	1.005	17,035.71	18,000

P14 = an activity selling screenings (damaged corn and foreign material separated by screening). Each unit of activity is 100 lb.

P15 = a corn transfer activity. The unit is one Winchester bushel of No. 2 corn.

P16 = an activity which transfers weight in excess of the minimum test weight from R01 (test weight equality) to the legal corn transfer row (R07).

P17 = a corn selling activity. The unit is one legal bushel.

Restraints

R01 = a restraint on test weight. The unit is pounds.

R02 = a restraint on moisture. The unit is pounds.

R03 = a restraint on damage. The unit is pounds.

R04 = a restraint on foreign material. The unit is pounds.

R05 = a transfer row for screenings. The unit is pounds.

R06 = an equality restraint placed on the activities to insure that the number of Winchester bushels being processed is equal to the number sold.

R07 = a corn grain transfer row. The unit is one legal bushel meeting No. 2 corn specifications.

Special Features of the Model

1. The objective function (C row) elements for the grain supplying activities (P01–P11) which are in units of Winchester bushels must be adjusted from legal bushel prices. The adjustment is made by dividing the Winchester bushel test weight by 56 and multiplying by the discounted price (the price for one legal bushel of No. 2 corn less discounts).
2. The moisture, damage, and foreign material percentages are converted to pounds, because percentages in Table 13.1 lack a common base due to test weight variations among bins.

3. Restraints R01 and R06 and activities P15 and P16 are formulated to insure that at least the minimum test weight per Winchester bushel is met. Row R06 is provided to insure that the number of measured bushels sold does not exceed the number processed (blended) for sale, and activity P15 provides for sale of one Winchester bushel which meets exactly the minimum test weight specification. Restraint R01 is necessary because all weight entering the blending process must be sold except for losses of moisture and dry matter in drying. Because activity P15 alone (excluding P16 from the model) provides for sale of corn weighing exactly 54 pounds per Winchester bushel, any weight in excess of the 54-pound test weight is permitted to be transferred for sale via activity P16. Thus, P15 and P16 together permit corn exceeding the minimum test weight to be blended and sold.
4. Row R07 appears in the model to facilitate converting and reporting the number of Winchester bushels processed into legal bushels. The P15-R12 coefficient equals 54/56 bushels or .964186 legal bushels.
5. An alternate approach to the blending problem is presented in Model 13.4. Both approaches lead to the same result upon optimization. The choice between them is a matter of convenience and the relative ease with which the user finds he can grasp the logic of each. Model 13.4 is presented because it requires less conversion of the grade specifications of the corn used for blending. The approach in Model 13.4 requires the definition of activity units in terms of 56 pounds (see restraint R01). This permits coefficients specifying the amount of moisture (R03) and foreign material (R04) for each activity (quality of grain) to be expressed in percentages, because the activities have a common base (percent of 56 pounds). Likewise, the objective function elements are not converted to the Winchester bushel but are expressed as the discounted price per legal bushel, because each of the supplying activities P01–P11 is defined in legal bushels. However, the elements in row R06 must be expressed in terms of the number of Winchester bushels per legal bushel, which equals 56 divided by the test weight per Winchester bushel in each activity.

Interpretation and Comparison of Results

1. The values of the programs resulting from optimization of Models 13.3 and 13.4 are $5420.40 and $5420.49 respectively. The difference between the two values can be attributed to rounding error.
2. The number of Winchester bushels and legal bushels entering the

MODEL 13.3

	Row Type	B Column	P01	P02	P03	P04	P05	P06	P07	P08	P09	P10	P11	P12	P13	P14	P15	P16	P17
C	N		-1.002406	-1.021426	-.92857	-1.021426	-.9483	-.974457	-1.002406	-.937856	-.974813	-.847046	-.951152	-.001		1.37			1.04
R01	E	0.0	54.5	55	52.0	55	55.9	53.5	54.5	52	53	53	53	-.54	-2.8242		-54.0	-56	
R02	L		8.5565	8.25	7.02	8.085	8.1055	8.56	8.5565	7.8	8.109	7.579	7.155	-.0837	-2.8242		-8.37	-8.68	
R03	L		1.635	.77	2.184	1.43	8.3291	1.605	1.635	1.976	1.06	4.187	4.717	-.027	-.0135		-2.7	-2.80	
R04	L		1.417	1.595	2.236	1.595	3.0186	1.498	1.4715	2.08	.901	5.088	1.59	-.54	-.27		-1.62	-1.68	
R05	L													-.54		100			
R06	E		1	1	1	1	1	1	1	1	1	1	1	-.01	-.055		-1		
R07	L																-.964286	-1	1
UP BND1			18,000	18,000	18,000	18,000	18,000	18,000	18,000	18,000	18,000	18,000	18,000						

MODEL 13.4

	Row Type	B Column	P01	P02	P03	P04	P05	P06	P07	P08	P09	P10	P11	P12	P13	P14	P15	P16	P17
C	N		-1.03	-1.04	-1.00	-1.04	-.95	-1.02	-1.03	-1.01	-1.03	-.895	-1.005	-.001037	-.0252	1.37			1.04
R01	E		56	56	56	56	56	56	56	56	56	56	56	-.56	-2.9288		-56	-56	
R02	L		15.7	15	13.5	14.7	14.5	16	15.7	15	15.3	14.3	13.5	-.1551	-5.0		-15.5	-15.5	
R03	L		3.0	1.4	4.2	2.6	14.9	3	3.0	3.8	2.0	7.9	8.9	-.05	-.025		-5.0	-5.0	
R04	L		2.6	2.9	4.3	2.9	5.4	2.8	2.7	4.0	1.7	9.6	3.0	-1.0	-.05		-3.0	-3.0	
R05	L													-.56		100			
R06	E		1.02752	1.01818	1.07692	1.01818	1.00178	1.04673	1.02752	1.07692	1.05660	1.05660	1.05660	-.0103704	-.057037		-1.03704		
R07	L																-1.	-1	1
UP BND1			17,517.87	17,678.57	16,714.29	17,678.57	17,967.86	17,196.43	17,517.87	16,714.29	17,035.71	17,035.71	17,035.71						

blends are specified by the level of activities P01–P11 and are given in Table 13.2. The number of bushels in the blend resulting from Model 13.3 is 164299.02 and from Model 13.4 is 164299.41.

3. The screening activities (P12 in each model) resulted in 766.45 hundredweight of screenings being sold in both cases.
4. The shadow prices or income penalties for each activity (see Table 13.3) indicate the change in the value of the program that would result from an increase in the level of the activity by one unit. Thus, adding a bushel of P01 to the blend would increase income by $.029, while forcing an additional bushel of P08 into the mix would decrease income by $.007. The positive shadow price on P01 indicates that having more corn with bin 1 specifications would increase the value of the blend. However, the reader should recall that the amount of P01 available is bounded at 18,000 Winchester bushels.
5. As discussed before, the P16 activity is provided in the model for transfer of weight in excess of the 54 pounds per Winchester bushel. Because the average of the eleven bins does not equal the minimum test weight specification for No. 2 corn, some corn of low test weight is not included in the optimum blend and remains in bins 3, 8, and 11. The shadow price on activity P16 of both models indicates the sacrifice in the value of the program which results from forcing a blend to exceed the minimum total test weight.

TABLE 13.2: Activity Levels of Models 13.3 and 13.4

Activity	Level	
	Model 13.3	Model 13.4
P01	18,000.00	17,517.87
P02	18,000.00	17,678.57
P03	15,396.53	14,297.62
P04	18,000.00	17,678.57
P05	18,000.00	17,967.86
P06	18,000.00	17,196.43
P07	18,000.00	17,517.87
P08	0.00	0.00
P09	18,000.00	17,035.71
P10	18,000.00	17,035.71
P11	12,406.95	11,742.36
P12	141,935.57	136,866.89
P13	0.00	0.00
P14	766.45	766.45
P15	170,384.17	164,299.41
P16	0.00	0.00
P17	164,299.02	164,299.41

TABLE 13.3: Income Penalties for Models 13.3 and 13.4

Activity	Income Penalty	
	Model 13.3	Model 13.4
P01	.02823	.02900
P02	.03287	.03346
P03	.00000	.00000
P04	.02797	.02847
P05	.07320	.07334
P06	.01892	.01979
P07	.02786	.02861
P08	—.00666	—.00717
P09	.00829	.00874
P10	.08380	.08854
P11	.00000	.00000
P12	.00000	.00000
P13	—.07382	—.07830
P14	.00000	.00000
P15	.00000	.00000
P16	—.99541	—.99531
P17	.00000	.00000

6. The shadow prices for the restraints have not been presented because they provide little information of value.
7. Although P16 does not enter the solution, this activity should be provided whenever it is anticipated that (a) the corn available for blending may exceed the minimum test weight (i.e., if only bins one to seven were available for blending) and/or (b) the quantity to be blended is restricted to a specified level.

MODEL 13.5: *Optimizing the Mix of Livestock Production Activities and Ration Components Simultaneously*

Explanation

The conventional approach to planning livestock programs specifies the mix of feed required per unit of livestock activity. In some cases several livestock activities, each of which provides a nutrient mix from different sources of materials, may be included in the model. Detailed studies of optimum rations are then approached as separate problems. The nutrients required are specified, and a range of materials from which these requirements can be met are defined. The optimizing process becomes a search for the mix of materials which will minimize the cost of meeting the requirements. This model illustrates how the two approaches can be integrated. Both the nutritive components which underly so-called feed requirements and those that are supplied by feed materials are specified in detail.

Maximizing the objective function defined in terms of net income specifies the least-cost ration, because the former cannot be achieved in the absence of the latter.

Activities

P01 = a corn growing activity. The unit of activity is one acre.
P02 = a CCOM growing activity. The unit of activity is four acres.
P03 = a CSBCO growing and a soybean harvesting and selling activity. The unit of activity is four acres.
P04 = a corn harvesting activity. The unit of activity is one acre.
P05 = an oats harvesting activity. The unit of activity is one acre.
P06 = a hay harvesting activity. The unit of activity is one acre.
P07 = an activity which farrows pigs and feeds them to 40 lb. and sells the sow. The unit of activity is one sow and two litters of 7.5 pigs each.
P08 = feeding pigs from 40 lb. to 80 lb. The unit of activity is one animal.
P09 = feeding pigs from 80 lb. to 130 lb. The unit of activity is one animal.
P10 = feeding hogs from 130 lb. to 220 lb. The unit of activity is one hog.
P11 = selling hogs at 220 lb. The unit of activity is one hog.
P12 = an activity which feeds corn produced on the farm or purchased to pigs weighing 40–80 lb. The unit of activity is one bushel.
P13 = an activity which feeds oats produced on the farm or purchased to pigs weighing 40–80 lb. The unit of activity is one bushel.
P14 = an activity which feeds hay produced on the farm or purchased to pigs weighing 40–80 lb. The unit of activity is one pound.
P15 = an activity which purchases and feeds sorghum to pigs weighing 40–80 lb. The unit of activity is one pound.
P16 = an activity purchasing and feeding soybean meal (45.8% crude protein) to pigs weighing 40–80 lb. The unit of activity is one pound.
P17 = an activity purchasing feeding meat and bone meal (50.6% crude protein) to pigs weighing 40–80 lb. The unit of activity is one pound.
P18 = an activity purchasing a vitamin premix that can be fed to pigs weighing 40–80 lb. The unit of activity is one pound.
P19 = an activity purchasing and feeding dicalcium phosphate to pigs weighing 40–80 lb. The unit of activity is one pound.
P20 = an activity transferring cystine to meet up to 40% of the methionine requirement. The unit of activity is one pound.
P21 = an activity purchasing methionine that can be fed to pigs weighing 40–80 lb. The unit of activity is one pound of actual methionine.

P22–P31 = the same as P12–P21 above except that the feed is being fed to pigs weighing 80–130 lb.
P32–P41 = the same as P12–P21 above except that the feed is being fed to hogs weighing 130–220 lb.
P42 = a corn selling activity. The unit of activity is one bushel.
P43 = a corn purchasing activity. The unit of activity is one bushel.
P44 = a hay selling activity. The unit of activity is one ton.
P45 = an oats selling activity. The unit of activity is one bushel.
P46 = an oats buying activity. The unit of activity is one bushel.
P47 = a labor hiring activity. The unit of activity is one hour.

Restraints

R01 = a land restraint. The B column entry is acres.
R02 = a labor restraint. The B column entry is hours.
R03 = a standing corn transfer row. The transfer unit is one acre.
R04 = a standing oats transfer row. The transfer unit is one acre.
R05 = a standing meadow transfer row. The unit of transfer is one acre.
R06 = a corn grain transfer row. The unit of transfer is one bushel.
R07 = an oats grain transfer row. The unit of transfer is one bushel.
R08 = a hay transfer row. The unit of transfer is one pound.
R09 = a capacity restraint on the number of sows that can be farrowed. The B column entry is number of sows.
R10 = a transfer row for 40 lb. pigs. The transfer unit is one pig.
R11 = a crude protein transfer. The unit of transfer is one pound.
R12 = a digestible energy transfer. The unit of transfer is 10,000 Kcal.
R13 = a calcium transfer. The transfer unit is one pound.
R14 = a phosphorus transfer. The transfer unit is one pound.
R15 = a vitamin A transfer. The transfer unit is one International Unit.
R16 = a vitamin D transfer row. The transfer unit is one International Unit.
R17 = a thiamine transfer row. The transfer unit is one milligram.
R18 = a riboflavin transfer row. The unit of transfer is one milligram.
R19 = a niacin transfer row. The unit of transfer is one milligram.
R20 = a pantothenic acid transfer row. The unit of transfer is one milligram.
R21 = a methionine transfer row. The unit of transfer is one pound.
R22 = a tryptophan transfer row. The unit of transfer is one pound.
R23 = a lysine transfer row. The unit of transfer is one pound.

MODEL 13.5

	B	P01	P02	P03	P04	P05	P06	P07	P08	P09	P10	P11	P12	P13	P14	P15	P16	P17	P18	P19	P20	P21	P22	P23	P24	
C N		-32.10	-73.86	32.89	-5.00	-4.50	-14.00	-26.42	-.550	.488	-.778	43.00				-.02	-.0525	-.0535	-.60	-.054		-.75				C
R01 L	480	1	4	4																						R01
R02 L	2,800	2.75	6.55	10.25	1.5	1.0	5.25	13.94	.483	.483	.724															R02
R03 L		-1	-2	-2	1																					R03
R04 L			-1	-1		1																				R04
R05 L			-1				1																			R05
R06 L					-100			61					1										1			R06
R07 L						-80								1										1		R07
R08 L							-7000								1										1	R08
R09 L	30							1																		R09
R10 L								-15	1																	R10
R11 G									-17.973				4.928	3.776	.167	.11	.458	.506				1				R11
R12 G									-16.8300				9.63704	.4160	.062818	.156954	.190818	.129954								R12
R13 G									-.674				.0168	.032	.0124	.0004	.0032	.1057		.222						R13
R14 G									-.562				.1512	.1120	.0028	.0029	.0067	.0507		.1790						R14
R15 G									-66,000										480,000							R15
R16 G									-10,200										48,000							R16
R17 G									-57.0				101.92	90.24		1.773	3.0	.5								R17
R18 G									-132.0				28.0	23.26		.545	1.5	2.0	300							R18
R19 G									-714.0					229.76			12.182	21.727	1600							R19
R20 G									-561.0				127.29	187.52		5.182	6.591	1.682	1200							R20
R21 G									-.5808				.0952	.0576		.0009	.006	.007			1	1				R21
R22 G									-.1373				.0504	.0576		.0009	.006	.002								R22
R23 G									-.8448				.1008	.1152		.0027	.0291	.035								R23
R24 G									0				.0504	.0576		.0018	.0067	.006			-1					R24
R25 G									-561.0										2000							R25
R26 L									-5.28				1.12	3.52	.258	.02	.06	.022								R26
R27 L									-.2323												1					R27
R28 L									-1	1																R28
R29 L										-1	1															R29
R30 L								1			-1	1														R30
R31 G										-23.127													4.928	3.776	.167	R31

R32 G										24.7500													9.63704	.4160	.062818	R32
R33 G										-.826													.0168	.032	.0124	R33
R34 G										-.661													.1512	.112	.0028	R34
R35 G										-97,500																R35
R36 G										-9,360																R36
R37 G										-84													101.92	90.24		R37
R38 G										-165													28.0	23.26		R38
R39 G										-750														229.76		R39
R40 G										-825													127.29	187.52		R40
R41 G										-.7088													.0952	.0576		R41
R42 G										-.1890													.0504	.0576		R42
R43 G										-1.1025													.1008	.1152		R43
R44 G																							.0504	.0576		R44
R45 G										-825																R45
R46 L										-7.88													1.12	3.52	.258	R46
R47 L										-.2835																R47
R48 G											-45.099															R48
R49 G											-51.9750															R49
R50 G											-1.734															R50
R51 G											-1.388															R51
R52 G											-204,750															R52
R53 G											-19,665															R53
R54 G											-175.5															R54
R55 G											-346.5															R55
R56 G											-1575															R56
R57 G											-1,732.5															R57
R58 G											-1.283															R58
R59 G											-.3209															R59
R60 G											-1.9251															R60
R61 G											0															R61
R62 G											-1732.5															R62
R63 L											-16.6															R63
R64 L											-.5132															R64
R65 G	14											1														R65

MODEL 13.5 (continued)

	P25	P26	P27	P28	P29	P30	P31	P32	P33	P34	P35	P36	P37	P38	P39	P40	P41	P42	P43	P44	P45	P46	P47	
C	-.02	-.0525	-.0535	-.60	-.054		-.75				-.02	-.0535	-.60	-.054	-.054		-.75	1.15	-1.20	20.50	.60	-.64	-2.00	C
R01																								R01
R02																							-1	R02
R03																								R03
R04																								R04
R05																								R05
R06								1										1	-1					R06
R07									1												1	-1		R07
R08										1										2,000				R08
R09																								R09
R10																								R10
R11																								R11
R12																								R12
R13																								R13
R14																								R14
R15																								R15
R16																								R16
R17																								R17
R18																								R18
R19																								R19
R20																								R20
R21																								R21
R22																								R22
R23																								R23
R24																								R24
R25																								R25
R26																								R26
R27																								R27
R28																								R28
R29																								R29
R30																								R30
R31	.11	.458	.506				1																	R31
R32	.156954	.190818	.129954																					

R33	.0004	.0032	.1057		.222																			R33
R34	.0029	.0067	.0507		.179																			R34
R35				480,000																				R35
R36				48,000																				R36
R37	1.773	3.0	.5																					R37
R38	.545	1.5	2.0	300																				R38
R39		12.182	21.727	1,600																				R39
R40	5.182	6.591	1.682	1,200																				R40
R41	.0009	.006	.007			1	1																	R41
R42	.0009	.006	.002																					R42
R43	.0027	.0291	.035																					R43
R44	.0018	.0067	.006			-1																		R44
R45				2,000																				R45
R46	.02	.006	.022																					R46
R47						1																		R47
R48								4.928	3.776	.167	.11	.458	.506				1							R48
R49								9.63704	.4160	.062818	.156954	.190818	.129954											R49
R50								.0168	.032	.0124	.0004	.0032	.1057		.222									R50
R51								.1512	.112	.0028	.0029	.0067	.0507		.179									R51
R52														480,000										R52
R53														48,000										R53
R54								101.92	90.24		1.773	3.0	.5											R54
R55								28.0	23.26		.545	1.5	2.0	300										R55
R56									229.76			12.182	21.727	1,600										R56
R57								127.29	187.52		5.182	6.591	1.682	1,200										R57
R58								.0952	.0576		.0009	.006	.007			1	1							R58
R59								.0504	.0576		.0009	.006	.002											R59
R60								.1008	.1152		.0027	.0291	.035											R60
R61								.0504	.0576		.0018	.067	.006			-1								R61
R62														2,000										R62
R63								1.12	3.52	.258	.02	.006	.022											R63
R64																1								R64
R65																								R65

R24 = a cystine transfer row. The unit of transfer is one pound.
R25 = a vitamin B12 transfer row. The unit of transfer is one microgram.
R26 = a crude fiber transfer row. The unit of transfer is one pound.
R27 = cystine to methionine transfer row. The unit of transfer is one pound.
R28 = an 80 lb. pig transfer row. The unit of transfer is one pig.
R29 = a 130 lb. pig transfer row. The unit of transfer is one pig.
R30 = a 220 lb. pig transfer row. The unit of transfer is one pig.
R31–R47 = the same as R11–R27.
R48–R64 = the same as R11–R27.
R65 = minimum restraint on the number of hogs that can be sold. B column entry is number of hogs.

Points to Observe

1. Because the level of hog production should be permitted to emerge from the optimization process, nutrient requirements cannot be placed in the B column as in an ordinary feed mix problem. Entering the nutrient requirements for each activity unit (i.e., one hog) in the internal portion of the model permits the *total* nutrient requirements to vary with the level of hog production. It should be noted that coefficients specifying nutrient requirements in this case are negative. This is necessary because whenever the feeding activities (P08, P09, P10) enter the plan, they are creating nutrient requirements (much the same as corn growing [P01] creates one acre of standing corn) which must be met by the feedstuffs offered in the program. These negative nutrient coefficients can also be thought of as becoming positive requirements when transferred to the B column side of the equation.
2. Where there are infeasibilities in the model or faulty coefficients which make feeding hogs unprofitable, there typically will be no ready clues as to the nature of the difficulty in the output report. Row R65 has been included to serve as a diagnostic tool. Once the individual is satisfied that his model does not contain major errors, this row can be eliminated from the model.
3. Because of the wide variation in the size of the coefficients in the model (e.g., 480,000 and .0009), cycling during an attempt to optimize becomes a hazard. Forestalling this problem requires modification of the control deck. The following set of control program statements is needed for optimization of the example model and

MODIFIED STANDARD PROCEDURE

Job Control Language and Control Program Cards

```
JOB CONTROL LANGUAGE CARDS


               PROGRAM
               INITIALZ
               MOVE(XDATA,'ECON430')
               MOVE(XPBNAME,'PBFILE')
               MVADR(XMAJERR,UNB)
               MVADR(XDONFS,NOF)
               CONVERT
               SETUP('MAX','SCALE')
               MOVE(XRHS,'B')
               MOVE(XOBJ,'C')
               XEPS=0.001
               XFREQINV=5
               XINVERT=1
               CRASH
               PRIMAL
               SAVE
               SOLUTION
               XEPS=0.00
               RESTORE
               PRIMAL
               SOLUTION
               EXIT
 NOF           TRACE
 UNB           EXIT
               PEND
/*


JOB CONTROL LANGUAGE CARDS


 NAME          ECON430


 ENDATA
/*
```

is recommended for problems that have large variations in size of numbers.

4. In formulating the nutrition requirements for activities P08, P09, and P10, a set of assumptions about the average daily rate of gain is required. These assumptions are given in Table 13.4. From these average daily gain (ADG) figures it is possible to calculate the number of days required for specified amounts of gain.

TABLE 13.4: Rates of Gain Assumed in Model 13.5

Activity	Assumed Average Daily Gain	Length of Period	Assumed Feed Requirements/ Day
P08 (40–80 lb. pig)	1.32	30 days	3.52
P09 (80–130 lb. pig)	1.65	30 days	5.25
P10 (130–220 lb. pig)	1.98	45 days	7.13

The length of each production period is also given in Table 13.4. To arrive at coefficients for a period, the number of days in a period are multiplied by the animals' daily nutrient requirements. For example, the protein requirement in P08 was calculated as follows: (40 lb gain/1.32 ADG) (.5991 lb protein required per day) = 17.973 lb protein.

Where daily nutrient requirements are not available, it is necessary to work from average daily feed intake. Estimates are shown in Table 13.4. These levels of feed consumption are multiplied by the number of days in the period to give the total feed requirements for one pig. An example would be the methionine requirement in P08: (3.52 lb feed/day) (30 days) (.55%) = .5808 lb.

Results

Optimization of models patterned after the above illustration permits the planner to judge which crop and livestock systems are the most profitable and also provides insight concerning optimum rations.

When Model 13.5 was optimized all 480 acres of land were used in corn production (P01). All of the operator's labor was required, and it was necessary to hire an additional 419 hours. Hog production was carried on at the maximum 60 litters permitted. The optimum rations reported were as follows:

RATION 1 (40–80 lb.)

Corn	50.6%	
Sorghum	26.8%	
Bean meal	16.6%	This ration contained
Meat and bone meal	5.6%	17.9% crude protein and cost $59.82/ton
Vitamin premix	.3%	for 22.52 tons.
Commercial methionine	.19%	

RATION 2 (80–130 lb.)

Corn	79.07%	
Bean meal	15.87%	This ration contained

Meat and bone meal	4.74%	16.6% crude protein and cost $58.16/ton for 32.28 tons.
Vitamin premix	.29%	
Commercial methionine	.03%	

RATION 3 (130–220 lb.)

Corn	79.75%	This ration contained 14.8% crude protein and cost $54.50/ton for 68.46 tons.
Sorghum	4.54%	
Bean meal	10.58%	
Meat and bone meal	4.83%	
Vitamin premix	.28%	
Commercial methionine	.02%	

Corn not fed to hogs was sold at $1.15/bu.

APPENDIX

This appendix is included for the student with an understanding of elementary matrix algebra. Its purpose is to demonstrate the kinship between the method for optimizing explained in Chapter 2 and the process of inverting a matrix. The latter task may be approached using several alternative methods. Here we describe one method which provides the most insight into the manipulations performed in the simplex method.

The following pattern of symbolic representation is used in this discussion:

a. Capital letters represent matrices.
b. Lower case letters represent elements within matrices.
c. The prime mark represents the transposition of the matrix.
d. A capital letter with an exponent of a negative one (—1) represents the inversion of the matrix.
e. I represents an identity matrix.
f. In citing matrices the subscript o in this presentation indicates a matrix none of whose elements is in the basis. The subscript s refers to a matrix whose elements are in the basis.
g. Nonsubscripted matrices are defined in the text.

First we organize the coefficients from the problem in the usual way. Note that the problem is the same in every detail as the one presented and optimized in Chapter 2. The disposal activities must be included.

Now we begin a process of partitioning the matrix and rearranging the matrices thus found. The first step is to form four matrices as follows:

$$S \quad \begin{matrix} [X_o' & X_s'] \\ [P_o & P_s\ ] \\ [C_o' & C_s'] \end{matrix} \tag{A.1}$$

Note that we have formed four matrices that relate to the simplex formulation as follows:

a. The S matrix is the B column or the supply matrix.
b. The X matrix encompasses the variables (activities), both real and disposal.
c. The P matrix includes all of the production coefficients and the coefficients (the ones and zeros) for the disposal section of the original matrix.
d. The C matrix is the C row or objective function.

Next we manipulate the newly created matrices (the parts of the original matrix, each one of which can be and is treated as a single matrix) to describe two relationships: (a) one between the variables (the X matrix) times their production coefficients (the P matrix) and the restraint level or supply (the S matrix) and (b) another between the level of the activities (the X matrix) times the appropriate net price of each (the C matrix) and value of the program.

Matrices if conformable can be multiplied as implied in the relationships above, but one must take care to multiply in the proper order. The conformable requirement means (when two matrices are multiplied) that the number of columns of the first matrix must be equal to the number of rows of the second. In the example which follows, $B \times A$ results in two conformable matrices (the number of columns of the first matrix equals the number of rows of the second) whereas $A \times B$ is nonconformable. (Even if conformable $A \times B \neq B \times A$.)

$$A = \begin{bmatrix} a_{11} & a_{12} & a_{13} \\ a_{21} & a_{22} & a_{23} \\ a_{31} & a_{32} & a_{33} \end{bmatrix}$$

$$B = \begin{bmatrix} b_{11} & b_{12} & b_{13} \\ b_{21} & b_{22} & b_{23} \end{bmatrix}$$

The process involved in multiplying $B \times A$ can be illustrated as follows:

$$\begin{bmatrix} b_{11} & b_{12} & b_{13} \\ b_{21} & b_{22} & b_{23} \end{bmatrix} \begin{bmatrix} a_{11} & a_{12} & a_{13} \\ a_{21} & a_{22} & a_{23} \\ a_{31} & a_{32} & a_{33} \end{bmatrix} =$$

$$\begin{bmatrix} a_{11}b_{11}+a_{21}b_{12}+a_{31}b_{13}, & a_{12}b_{11}+a_{22}b_{12}+a_{32}b_{13}, & a_{13}b_{11}+a_{23}b_{12}+a_{33}b_{13} \\ a_{11}b_{21}+a_{21}b_{12}+a_{31}b_{23}, & a_{12}b_{21}+a_{22}b_{22}+a_{32}b_{23}, & a_{13}b_{21}+a_{23}b_{22}+a_{33}b_{23} \end{bmatrix}$$

Then, substituting numerical values and multiplying as above, we have:

$$\begin{bmatrix} 2 & 3 & 6 \\ 4 & 1 & 2 \end{bmatrix} \begin{bmatrix} 1 & 4 & 8 \\ 2 & 6 & 3 \\ 5 & 4 & 2 \end{bmatrix} = \begin{bmatrix} (1)(2) + (2)(3) + (5)(6) & (4)(2) + (6)(3) + (4)(6) & (8)(2) + (3)(3) + (2)(6) \\ (1)(4) + (2)(1) + (5)(2) & (4)(4) + (6)(1) + (4)(2) & (8)(4) + (3)(1) + (2)(2) \end{bmatrix} = \begin{bmatrix} 38 & 50 & 37 \\ 16 & 30 & 39 \end{bmatrix}$$

Matrix inversion is that process in matrix algebra which corresponds to division in conventional algebra. If we begin with the equation $ay = 10$, we can solve for y algebraically by dividing both sides of the equation by a.

Thus, $y = \frac{10}{a}$. But we can also multiply both sides of the equation by $\frac{1}{a}$ (the inverse of a) and achieve the same result:

$$\frac{1}{a}\, ay = \frac{1}{a}\, 10 \quad \text{or} \quad y = \frac{10}{a}$$

This latter process must be followed in matrix algebra because matrix division is not defined.

The process of inversion is explained below. Setting up for the inverse we have:

$$\left(\begin{array}{ccc|ccc} a_{11} & a_{12} & a_{13} & 1 & 0 & 0 \\ a_{21} & a_{22} & a_{23} & 0 & 1 & 0 \\ a_{31} & a_{32} & a_{33} & 0 & 0 & 1 \end{array}\right)$$

By addition, subtraction, multiplication, and division we reduce the (a_{ij}/I) matrix to (I/a_{ij}^{-1}) matrix. Where the a_{ij}'s are the elements of the P_s matrix and the a_{ij}^{-1}'s are the elements of the inverse matrix.

As an example, let us invert the matrix:

$$\begin{pmatrix} 1 & 1 & 0 \\ 2 & 6 & 0 \\ 18 & 36 & 1 \end{pmatrix} \tag{A.2}$$

We form an identity matrix to the right of the bar. The identity matrix is that matrix which when multiplied by the original matrix will result in the original matrix. Setting up in the form (a_{ij}/I) we have (A.3):[1]

[1] Roger L. Burford, *Statistics, A Computer Approach* (Charles E. Merrill Publishing Co., Columbus, Ohio. 1968).

$$\left(\begin{array}{ccc|ccc} 1 & 1 & 0 & 1 & 0 & 0 \\ 2 & 6 & 0 & 0 & 1 & 0 \\ 18 & 36 & 1 & 0 & 0 & 1 \end{array}\right) \quad \text{(A.3)}$$

To find the inverse we need to do the necessary calculations to reduce the matrix left of the bar to an identity matrix. In the process the identity matrix becomes the inverse. Finding the inverse is critical to the determination of the incoming vectors, calculations of the ratio to determine the outgoing row, calculation of the B column, and determination of the opportunity cost estimate. Proceeding, we (a) multiply the first row of (A.3) by 2 and subtract from the second row of (A.3) to derive the new second row in (A.4); (b) multiply the first row in (A.3) by 18 and subtract from the third row in (A.3) to derive the new third row in (A.4):

$$\left(\begin{array}{ccc|ccc} 1 & 1 & 0 & 1 & 0 & 0 \\ 0 & 4 & 0 & -2 & 1 & 0 \\ 0 & 18 & 1 & -18 & 0 & 1 \end{array}\right) \quad \text{(A.4)}$$

At this point the first column of the original matrix has the form of the first column of the identity matrix. We now proceed to manipulate the matrix so that the second column is in identity form.

a. Dividing row 2 in (A.4) by 4 we have row 2 in (A.5).
b. Multiplying row 2 in (A.4) by —1/4 and adding to row 1 in (A.4) we obtain the new row 1 in (A.5).
c. For row 3 in (A.5) we multiply row 2 in (A.4) by —18/4 and add to row 3:

$$\left(\begin{array}{ccc|ccc} 1 & 0 & 0 & 3/2 & -1/4 & 0 \\ 0 & 1 & 0 & -1/2 & 1/4 & 0 \\ 0 & 0 & 1 & -9 & -9/2 & 1 \end{array}\right) \quad \text{(A.5)}$$

Column 3 of the A matrix is already in identity form. Thus the inverse of A is the matrix to the right of the bar in (A.5). The reader can demonstrate to his own satisfaction that the inverse has been obtained by multiplying the inverse times the original matrix (that part of [A.3] to the left of the bar).

With knowledge of matrix partitioning multiplication and inversion at hand, we can determine the optimum plan.

The matrices are partitioned as follows for the first iteration:

$$P_o = \begin{bmatrix} 1 & 1 & 1 \\ 6 & 6 & 2 \\ 36 & 24 & 18 \end{bmatrix} \quad X_s = \begin{bmatrix} x_{\text{land}} \\ x_{\text{labor}} \\ x_{\text{capital}} \end{bmatrix}$$

$$P_s = \begin{bmatrix} 1 & 0 & 0 \\ 0 & 1 & 0 \\ 0 & 0 & 1 \end{bmatrix} \quad X_o = \begin{bmatrix} x_{corn} \\ x_{soybeans} \\ x_{oats} \end{bmatrix}$$

$$S = \begin{bmatrix} 12 \\ 48 \\ 360 \end{bmatrix}$$

To determine the optimum solution we calculate (a) the R matrix where $R = P_s^{-1}P_o$ (the inverse of the basis coefficients times the non-basis coefficients) and (b) matrix D (the returns less opportunity cost matrix) where $D = C_o' - C_s'\ R$. The largest positive value of the D matrix determines which of the X_o variables (activities) has the greatest potential to increase the value of the program.

Forming the P_s^{-1} we have:

$$\left(\begin{array}{ccc|ccc} 1 & 0 & 0 & 1 & 0 & 0 \\ 0 & 1 & 0 & 0 & 1 & 0 \\ 0 & 0 & 1 & 0 & 0 & 1 \end{array}\right)$$

Because the P_s matrix is already in the identity form, the identity matrix to the right of the bar is also the inverse of the P_s matrix (this situation exists only in the initial iteration).

Now to find the R matrix:

$$R = \begin{bmatrix} 1 & 0 & 0 \\ 0 & 1 & 0 \\ 0 & 0 & 1 \end{bmatrix} \begin{bmatrix} 1 & 1 & 1 \\ 6 & 6 & 2 \\ 36 & 24 & 18 \end{bmatrix} = \begin{bmatrix} 1 & 1 & 1 \\ 6 & 6 & 2 \\ 36 & 24 & 18 \end{bmatrix}$$

Observe that $R = P_o$ only in the initial iteration. Then

$$D = [40 \quad 30 \quad 20] - [0 \quad 0 \quad 0] \begin{bmatrix} 1 & 1 & 1 \\ 6 & 6 & 2 \\ 36 & 24 & 18 \end{bmatrix}$$

$$= [40 \quad 30 \quad 20] - [0 \quad 0 \quad 0]$$

$$= [40 \quad 30 \quad 20]$$

Note that the elements of $C_s'\ R$ (the opportunity costs) are zero for each of the activities since no production is yet implied.

Selecting the largest positive element from the D matrix, we see that X_{corn} is the incoming variable in the X_s matrix and its input coefficients enter the P_s matrix. We must now determine which of the resources is the most limiting and hence will leave the basis. Recall that $x_i r_i \leq s_i$. Let x_i correspond to level of activity entering the basis, r_i represent the corresponding element in the R matrix, and s_i represent the corresponding element in the S matrix. Determination of the ratio as shown here follows exactly the procedure described for the simplex solution in Chapter 2.

Thus,

$$X_{\text{land}} = \frac{S_{\text{land}}}{r_{\text{land}}} = \frac{12}{1} = 12$$

$$X_{\text{labor}} = \frac{S_{\text{labor}}}{r_{\text{labor}}} = \frac{48}{6} = 8$$

$$X_{\text{capital}} = \frac{S_{\text{capital}}}{r_{\text{capital}}} = \frac{360}{36} = 10$$

because r_{land}, r_{labor}, and r_{capital} are the coefficients of the R matrix column corresponding to the element of the largest positive element of the D matrix.

After the first iteration

$$S = \begin{bmatrix} 12 \\ 48 \\ 360 \end{bmatrix} - 8 \begin{bmatrix} 1 \\ 6 \\ 36 \end{bmatrix} = \begin{bmatrix} 4 \\ 0 \\ 72 \end{bmatrix}$$

Next we substitute the X_{corn} into the S and X_s matrices:

$$S = \begin{bmatrix} 4 \\ 8 \\ 72 \end{bmatrix} \text{ and } X_s = \begin{bmatrix} X_{\text{land}} \\ X_{\text{corn}} \\ X_{\text{capital}} \end{bmatrix}$$

To calculate the value of the program at the end of the first iteration the C_s' matrix is multiplied by the S matrix:

$$V = [0 \quad 40 \quad 0] \qquad \begin{bmatrix} 4 \\ 8 \\ 72 \end{bmatrix}$$

$$V = (4 \times 0) + (8 \times 40) + (72 \times 0)$$
$$V = \$320$$

Note that multiplication of the two matrices results in the scalar (one row $\times$ one column) matrix.

We now repeat the cycle for the second iteration. At the start we have:

$$P_s = \begin{bmatrix} 1 & 1 & 0 \\ 0 & 6 & 0 \\ 0 & 36 & 1 \end{bmatrix} X_s = \begin{bmatrix} X_{\text{land}} \\ X_{\text{corn}} \\ X_{\text{capital}} \end{bmatrix} P_o = \begin{bmatrix} 0 & 1 & 1 \\ 1 & 6 & 2 \\ 0 & 24 & 18 \end{bmatrix} X_o = \begin{bmatrix} X_{\text{labor}} \\ X_{\text{soybeans}} \\ X_{\text{oats}} \end{bmatrix}$$

$$C_o' = [0 \quad 30 \quad 20]$$
$$C_s' = [0 \quad 40 \quad 0\ ]$$

The reader at this point may wish to compare the partitioned matrices with those with which the first iteration began and note the alterations that have occurred. Note that for the second set of partitioned matrices X_{labor} and X_{corn} have reversed their positions in the X_o and X_s matrices. Note also that the input coefficients have changed their positions in the P_s and P_o matrices and net prices in the C matrices.

We proceed by inverting the P_s matrix:

$$P_s = \begin{bmatrix} 1 & 1 & 0 \\ 0 & 6 & 0 \\ 0 & 36 & 1 \end{bmatrix}$$

$$\left(\begin{array}{ccc|ccc} 1 & 1 & 0 & 1 & 0 & 0 \\ 0 & 6 & 0 & 0 & 1 & 0 \\ 0 & 36 & 1 & 0 & 0 & 1 \end{array}\right) \tag{A.6}$$

Because columns 1 and 3 in the matrix to the left of the bar are in the identity form, we need only alter column 2 to reach the identity form. We do so by

a. dividing row 2 by 6 to derive the new row 2;
b. multiplying row 2 by (—1/6) and adding to row 1 to form the new row 1; and
c. multiplying row 2 by (—6) and adding to row 3 to form the new row 3.

Now,

$$\left(\begin{array}{ccc|ccc} 1 & 0 & 0 & 1 & -1/6 & 0 \\ 0 & 1 & 0 & 0 & 1/6 & 0 \\ 0 & 0 & 1 & 0 & -6 & 1 \end{array}\right) \tag{A.7}$$

Therefore:

$$P_s^{-1} = \begin{bmatrix} 1 & -1/6 & 0 \\ 0 & 1/6 & 0 \\ 0 & -6 & 1 \end{bmatrix}$$

$$R = P_s^{-1}\,P_o = \begin{bmatrix} 1 & -1/6 & 0 \\ 0 & 1/6 & 0 \\ 0 & -6 & 1 \end{bmatrix} \begin{bmatrix} 0 & 1 & 1 \\ 1 & 6 & 2 \\ 0 & 24 & 18 \end{bmatrix}$$

$$R = \begin{bmatrix} -1/6 & 0 & 2/3 \\ 1/6 & 1 & 1/3 \\ -6 & -12 & 6 \end{bmatrix}$$

Calculating for the incoming vector, we have:

$$P = C_o' - C_s' R = [0 \quad 30 \quad 20] - [0 \quad 40 \quad 0] \begin{bmatrix} -1/6 & 0 & 2/3 \\ 1/6 & 1 & 1/3 \\ -6 & -12 & 6 \end{bmatrix}$$

$$= [-40/6 \quad -10 \quad 20/3]$$

Due to the presence of the positive 20/3 in the D matrix, we see that profits can be increased by introducing X_{oats} into the basis.

To find the most limiting ratio, we divide the elements in the S matrix by the appropriate elements in the R matrix.

$$\frac{S_{land}}{r_{land}} = 4 \div 2/3 = 6$$

$$\frac{S_{corn}}{r_{corn}} = 8 \div 1/3 = 24$$

$$\frac{S_{capital}}{r_{capital}} = 72 \div 6 = 12$$

The new S matrix becomes:

$$X_s = \begin{bmatrix} 4 \\ 8 \\ 72 \end{bmatrix} - 6 \begin{bmatrix} 2/3 \\ 1/3 \\ 6 \end{bmatrix} = \begin{bmatrix} 0 \\ 6 \\ 36 \end{bmatrix}$$

Substituting X_{oats} in the X_s for X_{land}, we obtain:

$$X_s = \begin{bmatrix} X_{oats} \\ X_{corn} \\ X_{capital} \end{bmatrix} = \begin{bmatrix} 6 \\ 6 \\ 36 \end{bmatrix} = S$$

The value of the program after the second iteration is:

$$V = [20 \quad 40 \quad 0] \begin{bmatrix} 6 \\ 6 \\ 36 \end{bmatrix}$$

$$= (20 \times 6) + (40 \times 6) + (0 \times 36)$$
$$= 120 + 240 = \$360$$

The reader should note that we have achieved the optimum solution following the second iteration. We proceed to a third iteration (a) to demonstrate that the value of the program cannot be increased by further substitutions in the basis and (b) to derive the shadow prices.

To begin the third iteration we again present the partitioned matrices with the alterations that have occurred as a result of the second iteration.

$$P_o = \begin{bmatrix} 0 & 1 & 1 \\ 1 & 6 & 0 \\ 0 & 24 & 0 \end{bmatrix} \quad \begin{matrix} C_o' = [0 & 30 & 0] \\ C_s' = [20 & 40 & 0] \end{matrix}$$

$$P_s = \begin{bmatrix} 1 & 1 & 0 \\ 2 & 6 & 0 \\ 18 & 36 & 1 \end{bmatrix}$$

$$S = \begin{bmatrix} 6 \\ 6 \\ 72 \end{bmatrix} X_s = \begin{bmatrix} X_{\text{oats}} \\ X_{\text{corn}} \\ X_{\text{capital}} \end{bmatrix} X_o = \begin{bmatrix} X_{\text{labor}} \\ X_{\text{soybeans}} \\ X_{\text{land}} \end{bmatrix}$$

Finding P_s^{-1} we have:

$$\left(\begin{array}{ccc|ccc} 1 & 1 & 0 & 1 & 0 & 0 \\ 2 & 6 & 0 & 0 & 1 & 0 \\ 18 & 36 & 1 & 0 & 0 & 1 \end{array}\right) \tag{A.8}$$

Multiplying row 1 by (—2) and adding to row 2 we form the new row 2. Multiplying row 1 by (—18) and adding to row 3 results in a new row 3:

$$\left(\begin{array}{ccc|ccc} 1 & 1 & 0 & 1 & 0 & 0 \\ 0 & 4 & 0 & -2 & 1 & 0 \\ 0 & 18 & 1 & -18 & 0 & 1 \end{array}\right) \tag{A.9}$$

We have now reduced column 1 to the identity form. To alter column 2, we proceed in a similar manner by (a) multiplying row 2 by (1/4) to derive a new row 2, (b) multiplying row 2 by (—1/4) and adding to row 1 to obtain the new row 1, and (c) multiplying row 2 by (—9/2) and adding to row 3 to form a new row 3.

As a result we have (A.10).

$$\left(\begin{array}{ccc|ccc} 1 & 0 & 0 & 3/2 & -1/4 & 0 \\ 0 & 1 & 0 & -1/2 & 1/4 & 0 \\ 0 & 0 & 1 & -9 & -9/2 & 1 \end{array}\right) \text{ and } P_s^{-1} = \begin{bmatrix} 3/2 & -1/4 & 0 \\ -1/2 & 1/4 & 0 \\ -9 & -9/2 & 1 \end{bmatrix} \tag{A.10}$$

Since $R = P_s^{-1} P_o$

$$R = \begin{bmatrix} 3/2 & -1/4 & 0 \\ -1/2 & 1/4 & 0 \\ -9 & -9/2 & 1 \end{bmatrix} \begin{bmatrix} 0 & 1 & 1 \\ 1 & 6 & 0 \\ 0 & 24 & 0 \end{bmatrix} = \begin{bmatrix} -1/4 & 0 & 3/2 \\ 1/4 & 1 & -1/2 \\ -9/2 & -12 & -9 \end{bmatrix}$$

and

$$C_s'R = [20 \quad 40 \quad 0] \begin{bmatrix} -1/4 & 0 & 3/2 \\ 1/4 & 1 & -1/2 \\ -9/2 & -12 & -9 \end{bmatrix}$$

$$= [+5 \quad 40 \quad 10]$$

Then,

$$\begin{aligned} D &= C_o' - C_s'R \\ &= [0 \quad 30 \quad 0] - [5 \quad 40 \quad 10] \\ &= [-5 \quad -10 \quad -10] \end{aligned}$$

Since all of the elements in the D matrix are negative, we know that the program has been optimized, because the opportunity cost of substituting any of the variables for those in the basis is greater than its net return.

From the final D matrix, we derive the marginal revenue products of labor and land and the income penalty for soybeans. The variables are listed below in the order in which they appear in the X_o' matrix:

a. the first element (—5) is the marginal revenue product of labor;
b. the second element (—10) is the income penalty for soybeans;
c. the third element (—10) is marginal revenue product of land.

INDEX